Chemical Analysis of Firearms, Ammunition, and Gunshot Residue

INTERNATIONAL FORENSIC SCIENCE AND INVESTIGATION SERIES

Series Editor: Max Houck

Firearms, the Law and Forensic Ballistics
T A Warlow
ISBN 9780748404322
1996

Scientific Examination of Documents: methods and techniques, 2nd edition
D Ellen
ISBN 9780748405800
1997

Forensic Investigation of Explosions
A Beveridge
ISBN 97807484 05657
1998

Forensic Examination of Human Hair
J Robertson
ISBN 9780748405671
1999

Forensic Examination of Fibres, 2nd edition
J Robertson and M Grieve
ISBN 9780748408160
1999

Forensic Examination of Glass and Paint: analysis and interpretation
B Caddy
ISBN 9780748405794
2001

Forensic Speaker Identification
P Rose
ISBN 9780415 27182 7
2002

Bitemark Evidence
B. J. Dorion
ISBN 9780824754143
2004

The Practice of Crime Scene Investigation
J Horswell
ISBN 9780748406098
2004

Fire Investigation
N Nic Daéid
ISBN 9780415248914
2004

Fingerprints and Other Ridge Skin Impressions
C Champod, C J Lennard, P Margot, and M Stoilovic
ISBN 9780415271752
2004

Firearms, the Law, and Forensic Ballistics, Second Edition
Tom Warlow
ISBN 9780415316019
2004

Forensic Computer Crime Investigation
Thomas A. Johnson
ISBN 9780824724351
2005

Analytical and Practical Aspects of Drug Testing in Hair
Pascal Kintz
ISBN 9780849364501
2006

Nonhuman DNA Typing: Theory and Casework Applications
Heather M Coyle
ISBN 9780824725938
2007

Chemical Analysis of Firearms, Ammunition, and Gunshot Residue
James Smyth Wallace
ISBN 9781420069662
2008

INTERNATIONAL FORENSIC SCIENCE
AND INVESTIGATION SERIES

Chemical Analysis of Firearms, Ammunition, and Gunshot Residue

James Smyth Wallace

CRC Press
Taylor & Francis Group
Boca Raton   London   New York

CRC Press is an imprint of the
Taylor & Francis Group, an **informa** business

CRC Press
Taylor & Francis Group
6000 Broken Sound Parkway NW, Suite 300
Boca Raton, FL 33487-2742

© 2008 by Taylor & Francis Group, LLC
CRC Press is an imprint of Taylor & Francis Group, an Informa business

No claim to original U.S. Government works
Printed in the United States of America on acid-free paper
10 9 8 7 6 5 4 3 2 1

International Standard Book Number-13: 978-1-4200-6966-2 (Hardcover)

This book contains information obtained from authentic and highly regarded sources Reasonable efforts have been made to publish reliable data and information, but the author and publisher cannot assume responsibility for the validity of all materials or the consequences of their use. The Authors and Publishers have attempted to trace the copyright holders of all material reproduced in this publication and apologize to copyright holders if permission to publish in this form has not been obtained. If any copyright material has not been acknowledged please write and let us know so we may rectify in any future reprint.

Except as permitted under U.S. Copyright Law, no part of this book may be reprinted, reproduced, transmitted, or utilized in any form by any electronic, mechanical, or other means, now known or hereafter invented, including photocopying, microfilming, and recording, or in any information storage or retrieval system, without written permission from the publishers.

For permission to photocopy or use material electronically from this work, please access www.copyright.com (http://www.copyright.com/) or contact the Copyright Clearance Center, Inc. (CCC) 222 Rosewood Drive, Danvers, MA 01923, 978-750-8400. CCC is a not-for-profit organization that provides licenses and registration for a variety of users. For organizations that have been granted a photocopy license by the CCC, a separate system of payment has been arranged.

Trademark Notice: Product or corporate names may be trademarks or registered trademarks, and are used only for identification and explanation without intent to infringe.

Library of Congress Cataloging-in-Publication Data

Wallace, James Smyth.
  Chemical analysis of firearms, ammunition, and gunshot residue / James Smyth Wallace.
    p. cm. -- (International forensic science and investigation ; 14)
  Includes bibliographical references and index.
  ISBN 978-1-4200-6966-2 (alk. paper)
  1. Chemistry, Forensic. 2. Firearms. 3. Ammunition. I. Title. II. Series.

HV8073.W334 2008
363.25'62--dc22   2008000780

Visit the Taylor & Francis Web site at
http://www.taylorandfrancis.com

and the CRC Press Web site at
http://www.crcpress.com

Dedication

To my first grandchild, Matthew

Contents

Preface	xi
Acknowledgments	xvii
About the Author	xix
Glossary	xxi

I INTRODUCTION

1 Definitions — 3

II HISTORICAL ASPECTS OF FIREARMS AND AMMUNITION

2 History of Gunpowder — 13

3 History of Ignition Systems — 15

4 History of Bullets — 19

5 History of Ammunition — 23

6 History of Firearms — 29

III CHEMICAL ASPECTS OF FIREARMS AND AMMUNITION

7 Cartridge Cases — 35

8 Primer Cups (Caps) — 39

9	Priming Compositions	41
10	Propellants	57
11	Projectiles	67
12	Complementary Ammunition Components	91
13	Caseless Ammunition	93
14	Blank Ammunition	95
15	Firearm Construction Materials	97

IV FIREARM DISCHARGE RESIDUE

16	Firearm Discharge Residue Detection Techniques	103
17	Properties of Firearm Discharge Residue	123

V EXPERIMENTAL

18	Objectives, Sampling Procedures, Instrumentation, and Conditions	137
19	Particle Classification Scheme	143
20	Casework-Related Tests	157
21	Analysis of Ammunition	183
22	Ammunition Containing Mercury	205
23	Lead-Free Ammunition	223

VI SUSPECT PROCESSING PROCEDURES

24 Firearm Discharge Residue Sampling   233

VII ORGANIC COMPONENTS OF FIREARM DISCHARGE RESIDUE

25 Sampling of Skin and Clothing Surfaces for Firearm Discharge Residue   241

26 Development of a Method for Organic Firearm Discharge Residue Detection   253

Conclusion   271

Index   277

Preface

There are numerous detailed books on firearms available for enthusiasts, the vast majority of the books concentrating on the physical aspects of firearms. Very little has been published on the chemical aspects of firearms and ammunition and what has been published is sparse and fragmented in the literature. One of the reasons for this is that manufacturers are reluctant to release in-depth details about their products for obvious commercial reasons.

The first part of the book is an attempt to amalgamate such chemical information as is available in the literature into one publication and also to summarize the history of firearms and ammunition that is of particular relevance to the development of modern firearms and ammunition (Chapter 2 through Chapter 15).

The remainder of the book details chemical aspects of forensic firearms casework with particular emphasis on the detection of gunshot residues (GSR)/firearm discharge residues (FDR)/cartridge discharge residues (CDR) on a suspect's skin and clothing surfaces. The development of an analytical method to routinely examine samples from terrorist suspects for both firearms and explosives residues is described.

Northern Ireland was subjected to a terrorist campaign for nearly 26 years (commonly referred to locally as "the troubles"). The violence is now ended, much to the relief of the overwhelming majority of residents, and the community is thriving. During the troubles the Northern Ireland Forensic Science Laboratory (NIFSL) experienced a large firearms caseload and this text is geared toward recording statistics gathered during this period and scientific methods developed to meet the demands of law enforcement and courts of law. The contents will be of interest to any forensic laboratory engaged in such work, particularly to forensic chemists, with little or no knowledge of firearms, who may be required to undertake chemical examinations related to firearms casework.

Sources include gun books, textbooks, gun magazines, scientific papers, technical reports, manufacturer's literature, newspaper articles, private communications, personal observations, and research conducted by myself.

The NIFSL has a turbulent history. It has been subjected to an armed raid by terrorists which resulted in a substantial number of firearms being stolen; an unsuccessful bombing attempt; a disastrous fire, the water and smoke damage from which destroyed the overwhelming majority of instrumentation;

and finally a large terrorist bomb which destroyed the laboratory and resulted in it being rebuilt in a different, more secure location. The forensic science staff members are civil servants and are totally independent of the police and army, but because of the nature of the work they were viewed as part of the so-called British war machine, and consequently the laboratory was targeted by some of the terrorist organizations. It may be of interest to some readers to briefly outline the main difference between firearms examination in a terrorist and a non-terrorist situation.

To explain the work of a laboratory dealing with a terrorist situation it is helpful to give a brief explanation of the background to the terrorist situation in Northern Ireland. The U.K. consists of England, Scotland, Wales, and Northern Ireland.

Northern Ireland is part of the United Kingdom but it is also part of the island of Ireland (Figure 0.1). The remainder of Ireland is a republic and is

Figure 0.1 Map of British Isles.

Preface

an independent sovereign nation with no political ties to any other country. Although it is the wish of the government of the republic to unite the whole of Ireland by political and peaceful means there are people, mostly Catholics, both north and south of Ireland, who wish to expel the British from Ireland by the use of the bomb and the bullet.

On the other hand there are people in Northern Ireland, mostly Protestants, who wish to maintain the position of Northern Ireland within the United Kingdom and who will use the bomb and bullet to further their own ends.

It must be stressed that among all these nationalists and unionists it is only a very small number who are involved in terrorism.

The end result of the civil unrest was numerous shootings and bombings both of members of the security forces and of people who are suspected of being associated with one side or the other. Much property was destroyed by bombings, and armed robberies to finance the various causes were common. All this resulted in a substantial financial burden on the state for a long period of time.

It is essential for any laboratory dealing with civil unrest to take the firm view that the law of the land is the only yardstick by which all criminal activity is measured and that assassinations, punishment shootings, shootings by the army and police, shootings of the army and police, and so forth are all shooting incidents and all demanded full and impartial investigation.

In 1969, brooding civil unrest unleashed the gunman, and the use of firearms in violent crime escalated and at the peak of the trouble the laboratory was dealing with approximately 2,800 firearms cases per year. Apart from the volume of casework the majority of cases were of a serious nature and many of the examinations were complex.

The equipment and methods employed by the firearms section do not differ from those used by other forensic laboratories. The section undertakes all forensic aspects of firearms examination both chemical and physical.

The section also provided a 24-hour, 7-day/week call out service to the security forces. Forensic staff members are available to attend scenes of crime which are large, controversial, or too difficult for the police or civilian scenes of crime officers to deal with. There were aspects of scenes of crime examination that most other laboratories do not experience: the possibility of booby traps or sniper attack; the scenes were often in hostile areas, so that the time to examine a scene was sometimes very limited (before a riot erupted or a sniper got organized); and many scenes extended over a large area, involved a large number of people and exhibits, and were of a controversial nature.

Police have been fired at while examining scenes and in one incident two policemen were killed by a booby-trapped shotgun when one attempted to check if the firearm was loaded. In another incident a policeman was killed while going from room to room in a house, one of the rooms having been booby trapped with an explosive device. At one stage, booby-trapped cars were common and were frequently involved in scenes of crime. On another occasion a booby-trapped rifle was received at the laboratory but fortunately the explosive device was discovered before the weapon was test fired. It became necessary to X-ray all relevant firearms and associated items before examination.

A particular firearm could be active over many years in terrorist hands and may or may not be recovered. A link report is a report that connects two or more shooting incidents by comparison macroscopy of spent cartridge cases and/or fired bullets, and we were frequently required to provide such reports for court purposes. This often involved a large amount of work. In one particular case 605 spent cartridge cases and 46 spent bullets had to be examined, and in addition to the 27 original reports a further 19 link reports were required for court purposes.

Link reports of this size would rarely be undertaken by other laboratories and are a direct consequence of terrorist activity. An interesting observation from doing link reports of this nature is that, for firearms used over a number of years, we could nearly always match spent cartridge cases whereas we were frequently unable to match all of the bullets. Terrorist weapons are generally neglected and fouling inside the barrel, storage under poor conditions, and so forth leads to rusting and wear inside the barrel which can substantially alter the striation markings on the bullet.

A further sinister aspect of the terrorist campaign was the use of heavy weaponry such as rocket launchers, mortars, and heavy machine guns. Items of this nature have been used to attack security force bases, police stations, and police vehicles. On more than one occasion army helicopters have been struck by large caliber machine gun fire.

Firearms and associated items, ammunition, spent bullets, and spent cartridge cases recovered from arms finds, scenes of crime, and so forth provide useful information of an intelligence nature.

The majority of the firearms recovered were rifles, revolvers, pistols, shotguns, and machine guns but items of a more unusual nature have been recovered including anti-tank rifles, rocket launchers, grenade launchers, flare pistols, air weapons, antiques, starting pistols, toy guns, riot guns, gas guns, humane killers, harpoon guns, industrial nail guns, line throwing guns, replicas, try guns, cross bows, zip guns, range finding binoculars, telescopic sights, tools, reloading and cleaning equipment, spare parts for a wide range of firearms, silencers, holsters, ammunition belts.

A wide range of items of a ballistics nature have been retained over the years and a large amount of information is stored on computer. This statistical database was a valuable aid to police investigating officers, and a similar intelligence framework operated for explosives and explosive devices.

Immediately after a shooting incident it is essential that the type of gun used and the history (if any) of the gun is established quickly. This may show which terrorist group last used the gun and the area in which the gun was last used, thereby giving the police an indication of where to look for the culprits and who were the likely suspects. Also in the case of motiveless shootings, if a link could be established with other shootings, then the organization involved may be identified. If a gun had a history and suspects were apprehended, then the history of the gun opened up a further line of questioning.

Another aspect of intelligence work was the possibility of tracing firearms or associated items back to the original supplier in another country and through examination of company records, receipts, and so forth trace the route of the gun to Ireland. This could lead to information about the purchaser and those involved in gun running and very occasionally resulted in prosecution of those involved. Such prosecutions have taken place in the United States and Australia.

Propaganda is a weapon that has been used very effectively by the terrorist. The database on firearms was frequently used by the police, army, politicians, and other official bodies to counter terrorist propaganda. Apart from armed robberies, terrorists financed their causes by donations from sympathizers both in Ireland and in other countries, and this is one example where facts are essential in order to inform supporters about the deeds and nature of the persons and organizations that they are financially encouraging.

To summarize, the main differences between firearms examination in a terrorist situation and a non-terrorist situation are that in a terrorist situation more cases tend to be of a serious nature, casework involves a wider variety of firearms and related items, more and larger link reports are required, there are different conditions and more difficulties with scene examination, and there is an intelligence gathering aspect to the work.

Acknowledgments

I am indebted to Mrs. Roseline E. Collins for her typing skills and to my wife, Edna, for her computer skills. My thanks also go to my colleagues at the Northern Ireland Forensic Science Laboratory whose dedication and enthusiasm were inspirational.

About the Author

Dr. Jim Wallace is a retired U.K. forensic scientist who worked in the firearms section of the Northern Ireland Forensic Science Laboratory for almost 25 years.

During this time he experienced many complex and controversial cases, the vast majority of which were terrorist-related incidents. His main interests include the chemical examinations relating to firearms casework, research and development work arising from same, crime scene examination, health and safety issues, quality assurance, suspect handling and processing, and contamination avoidance procedures both inside and outside the laboratory.

He is the author/co-author of 14 scientific papers and has contributed to a textbook on forensic science. He is a member of the Forensic Science Society and retains an active interest in forensic chemistry, particularly in the area of trace evidence detection.

Glossary

Chemical

ACN acetonitrile
DBP dibutylphthalate
DDNP diazodinitrophenol
DEGDN diethyleneglycoldinitrate
DMF dimethylformamide
DNB dinitrobenzene
DNT dinitrotoluene
DPA diphenylamine
EC ethylcentralite
EGDN ethyleneglycoldinitrate
IPA isopropyl alcohol
MC methylcentralite
MCE mixed cellulose esters
meDPA methylethyldiphenylamine
NB nitrobenzene
NC nitrocellulose
nDPA a nitro diphenylamine
NG nitroglycerine
PETN pentaerythritoltetranitrate
PTFE polytetrafluoroethylene
RDX cyclotrimethylenetrinitramine
TNT trinitrotoluene

Instrumental

FAAS flameless atomic absorption spectrophotometry

FTIR Fourier transfer infrared spectroscopy

GC/MS gas chromatography/mass spectrometry

GC/TEA gas chromatography/thermal energy analyzer

HPLC/PMDE high-performance liquid chromatography/pendant mercury drop electrode

NAA neutron activation analysis

SEM/EDX scanning electron microscopy/energy dispersive X-ray analysis

SPE solid phase extraction

Firearms/Ammunition

⊕ NATO specifications

.22 LR caliber .22 long rifle caliber

+P higher pressure ammunition

ACP automatic Colt pistol

AP armor piercing

Carbine a shorter length, lightweight rifle

FMJ full metal jacket

G gauge (bore)

H & K Heckler and Koch

HP bullet hollow point bullet

I bullet incendiary bullet

JHP bullet jacketed hollow point bullet

Jkt jacket

JSP bullet jacketed soft point bullet

K kurtz (short)

L long

Machine pistol   handgun-style pistol capable of automatic fire

Mag   magnum

NCNM   noncorrosive, nonmercuric

P   parabellum

Rem   Remington

Rev   revolver

RNL bullet   round-nosed lead bullet

S & W   Smith and Wesson

SMG   submachine gun

Spl   special

SWC bullet   semi-wad-cutter bullet

T bullet   tracer bullet

TMJ bullet   total metal-jacketed bullet

Win   Winchester

Miscellaneous

AFTE   Association of Firearm and Toolmark Examiners

ARDS   automatic residue detection system

CCI   Cascade Cartridge, Inc.

CDR   cartridge discharge residue

FBI   Federal Bureau of Investigation (U.S.)

FDR   firearm discharge residue

GSR   gunshot residue

IRA   Irish Republican Army

M level   major level

Mi level   minor level

NATO   North Atlantic Treaty Organization

NIFSL   Northern Ireland Forensic Science Laboratory

RPG   rocket-propelled grenade

SOCO scenes of crime officer

T level trace level

Notes

1. Firearms-related terms are defined in the Association of Firearm and Toolmark Examiners, *Glossary,* 3rd ed. (1994), published by Available Business Printing, Inc., 1519 South State Street, Chicago, IL 60605.
2. Ammunition details are given in H. P. White and B. D. Munhall, "Cartridge Headstamp Guide," published by H.P. White Laboratory, Bel Air, MD.
3. 7,000 grains (gn) = 1 pound (lb) = 453.59237 grams (g) = 16 ounces (oz.).
4. Firearms that were altered, damaged, or destroyed during the experimental work were all destined for disposal.
5. Persons referenced under "Private communications" cannot be identified for security reasons.
6. Due to a terrorist bomb attack on the NIFSL in September 1992, some material relating to the experimental work was destroyed or lost. Fortunately, the bulk of it was salvaged. However, some details were lost and these will be mentioned in the text. The missing paperwork included some of the references and I apologize to those whose work I have detailed but not referenced.

Introduction   I

Definitions 1

(a) Weapons

Handheld weapons preceded weapons designed to kill or incapacitate from a distance. Such weapons included wooden clubs and pointed sticks eventually leading to pointed, stone-tipped spears, and knives, daggers, and swords made from wood or stone.

The desire to propel some form of missile through the air to kill or injure a foe originated with primitive humans but this was probably not the initial objective. It is highly likely that the original reason was the necessity to hunt and kill dangerous animals for food and clothing, and for obvious reasons a weapon capable of killing from a safe distance would be highly desirable.

The first projectiles were probably stones and pointed wooden sticks which were initially thrown by hand. These developed through various stages including flint-tipped spears and arrows, eventually leading to propulsion using slings, throwing sticks, catapults, bows, and so forth, all of which gave the projectiles greater range, greater velocity, and consequently greater wounding power.

A major development in human armament was the discovery of metal and the ability to work metal, and this rapidly led to metal knives, daggers, and swords and metal-tipped spears and arrows. These were much superior to the wooden and stone weapons, and were used for many years until the development of a handheld weapon that surpassed all others and that had a profound effect on human history—*the firearm*.

(b) Firearms

It is probable that the word *firearm* originated from the flame produced at the muzzle end (muzzle flash) when a firearm is discharged; the muzzle is the front part where the bullet emerges from the barrel (Photograph 1.1).

The word *gun* is a widely accepted alternative name to firearm, although it is also used in many nonfirearm terms, for example, grease gun, spray gun, flame gun, nail gun, insecticide gun, paint gun, stun gun, etc. [A stun gun is

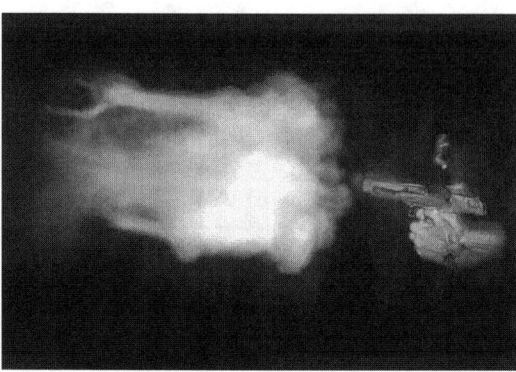

Photograph 1.1 Muzzle flash.

a weapon designed to disable a victim temporarily by delivering a nonlethal high-voltage electric shock. The gun needs to be in contact with the victim. A taser gun works in the same way but can be used at a distance, up to about 15 feet. It shoots small electrodes at the victim, thereby connecting the gun and victim through metal wires; 50,000 volts travel through the wires for 5 seconds. The gun has 18 watts of power output, which yields a very low amperage (0.00036 amps). Because amperage is very low, no serious or permanent injury is caused]. Heavy, large-caliber guns, as used in land/sea warfare, come under the category of artillery and are beyond the scope of this text. Small arms are firearms which can be carried by an individual.

Ballistics is the science of the performance of projectiles, relating to their trajectory, energy, velocity, range, penetration, and so forth. Exterior ballistics is concerned with the flight of the bullet after it leaves the muzzle of the gun. Interior ballistics is concerned with the primer ignition, the burning of the propellant powder, and the resulting internal pressures and torques as the bullet is forced through the barrel. Terminal ballistics is the study of the interaction of the projectile with the target.

A firearm is a tool designed to discharge lethal projectiles from a barrel toward selected targets. It is the means of aiming and discharging the projectile and imparting stability to it. In Northern Ireland the law defines a firearm as "a lethal barreled weapon of any description, from which any shot, bullet or other missile can be discharged."[1] This very broad definition does not mention the means of causing the shot, bullet, or other missile to be discharged, but this may be by compressed air, by gas (for example, carbon dioxide cylinders), by mechanical means (for example, a spring), or by the rapid burning of a propellant (gunpowder). Because a firearm operated by the burning of a propellant is by far the most common and potentially the most lethal, only this will be considered here. (Anything used to resemble a firearm in the commission of a crime may be treated as a real firearm in a criminal law trial.)

Definitions

Firearms can have different features such as the ability to carry more than one cartridge, to load and unload automatically, and to fire repeatedly on a single pressing of the trigger. These are refinements on the basic design. In its simplest form a firearm could be a crude metal tube with one end packed with some form of propellant which on ignition produces enough gas pressure to discharge a projectile or projectiles with sufficient energy to cause human death. In its most complicated form it might be a well-made and finely engineered tool capable of discharging and directing bullets on automatic fire up to a rate in the region of 1,500 rounds per minute over an accurate range of about 200 meters (1,500 rounds per minute is extreme, 600 to 800 rounds per minute is much more common and practical). Alternatively, it might be a high-powered, highly accurate sniper rifle equipped with telescopic sights and capable of killing a selected target at a distance of 1,000 meters or more.

Firearms are relatively cheap, readily produced, reliable, and deadly and find many uses, among which are warfare, sport, self-defense, hunting, law enforcement, and crime. It is the use of firearms in crime that demands the attention of the forensic scientist.

The most commonly used firearms in crime are pistols, revolvers, and rifles up to and including .455" caliber, and shotguns, the most popular of which is the "sawn-off" 12-bore caliber (Photograph 1.2 through Photograph 1.7 illustrate different types of firearms). This discussion deals with these weapons although submachine guns; machine guns; larger-caliber firearms; homemade firearms; air, spring, and gas guns; imitation and replica firearms are also encountered in crime, but to a much lesser extent. (A submachine gun is a lightweight machine gun that is hand held as compared to a machine gun that is bipod/tripod mounted for continuous fire.)

Pistols and revolvers are usually described as handguns, and rifles and shotguns as shoulder guns, as this is their normal mode of use.

A revolver is a type of pistol but for clarity it is better to describe it separately. A revolver is a single-barreled handgun with a revolving cylinder (multiple chambers), which holds a number of rounds of ammunition

Photograph 1.2 Machine gun.

Photograph 1.3 Submachine gun.

Photograph 1.4 Pistol.

Photograph 1.5 Revolver.

Definitions

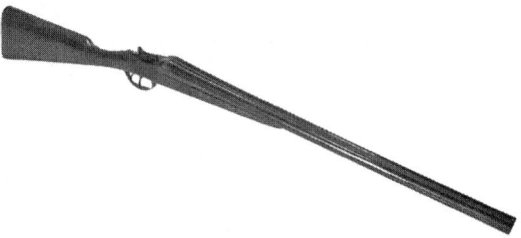

Photograph 1.6 Shotgun.

Photograph 1.7 Rifle.

(usually six). Each time the trigger is pulled, the cylinder is mechanically rotated so that each successive round of ammunition is placed in the firing position, that is, in line with the barrel. The spent cartridge cases are not ejected automatically but have to be removed manually.

A pistol is a single-barreled handgun in which the chamber is an integral part of the rear end of the barrel. A pistol can be either the single shot type, with manual or automatic ejection of the spent cartridge case, or much more commonly the self-loading type. In self-loading pistols a number of rounds of ammunition are loaded into a magazine, which is usually fitted into the handgrip of the weapon. Once the weapon is initially cocked and discharged a reloading mechanism, which is operated by the force of recoil or by gas pressure, extracts and ejects the spent cartridge case from the chamber and reloads the chamber with a live round of ammunition from the magazine. The process is repeated with each pull of the trigger until the ammunition is expended.

Rifles have a longer barrel than revolvers or pistols, and are usually more powerful and designed to shoot at targets at longer distances. Like pistols, they can be either the single shot type or the self-loading type. Rifles use various methods for the ejection of spent cartridge cases including lever, bolt, or pump action in manual operation, or recoil energy or gas pressure in automatic operation.

Shotguns can be either single or double barreled. The most common type is the design in which the barrel breaks forward on a hinge to expose the breech, into which live cartridges are manually inserted and from which spent cartridge cases are manually extracted. Some shotguns use pump action, recoil energy, or gas pressure reloading mechanisms (usually combat-type shotguns).

The barrel of a firearm is a tube made of iron or steel. The inner surface of the barrels of revolvers, pistols, and rifles contain a number of spiral grooves known as rifling. Hence they are known as rifled bore weapons. The rifling of the barrel grips the bullet and causes it to rotate, thereby preventing it from wobbling or turning over in flight.

The raised area between two grooves is called a land and the caliber of a firearm is based on the diameter of the bore (barrel) measured between two opposite lands.[2] This oversimplified definition of caliber gives a rough approximation of bullet diameter as the bullet is usually slightly bigger than the diameter of the bore. Caliber is usually given in inches or millimeters and common calibers for handguns are .22" (6 mm), .25" (6.35 mm), .32" (7.65 mm), .38"/.357" (9 mm), .45", .455", and for rifles are .22" (6 mm), .223", and .30" (7.62 mm). There are many other calibers in existence. In fact, the suitability of a round of ammunition for use with a particular firearm depends not only on the diameter of the bullet, but also on the length and design of the cartridge case. Caliber is often a misunderstood and confusing term as the following few examples illustrate: (a) .22 L and .22 LR—the same cartridge case is used in both, but the bullet weights are different. The bore diameter is 0.215 inches. (b) .308 Winchester—it was originally designed as a sporting cartridge in the United States but is also known as 7.62 × 51 NATO, adopted by NATO as the official military cartridge. The case length is 51 mm. (c) .45 ACP and .45 Auto Rim—the same bullet is used in each of these, but the design of the cartridge case differs. The .45 ACP has no rim and is designed for use in auto pistols, whereas the .45 Auto Rim has a rim to enable it to be used in a revolver. They are not interchangeable. (d) .380 Rev, .38 S&W, .38 Special, and .357 Magnum—in each of these the diameter of the bore is approximately 0.35 inches, but the cartridge case dimensions, bullet weights, and propellant charges are different, and each is designed for a different firearm, although some interchange is possible.

With very few exceptions the inner surface of a shotgun barrel is smooth; hence shotguns are called smooth bore weapons. The caliber of a shotgun is usually expressed as its bore or gauge; the most common are 12, 16, and 20 bore with the 12 bore by far the most popular. Bore refers to the number of lead balls of bore diameter that weigh 1 lb.[3] Smaller-diameter shotguns are usually described by the internal diameter of the barrel, for example, .410".

Definitions 9

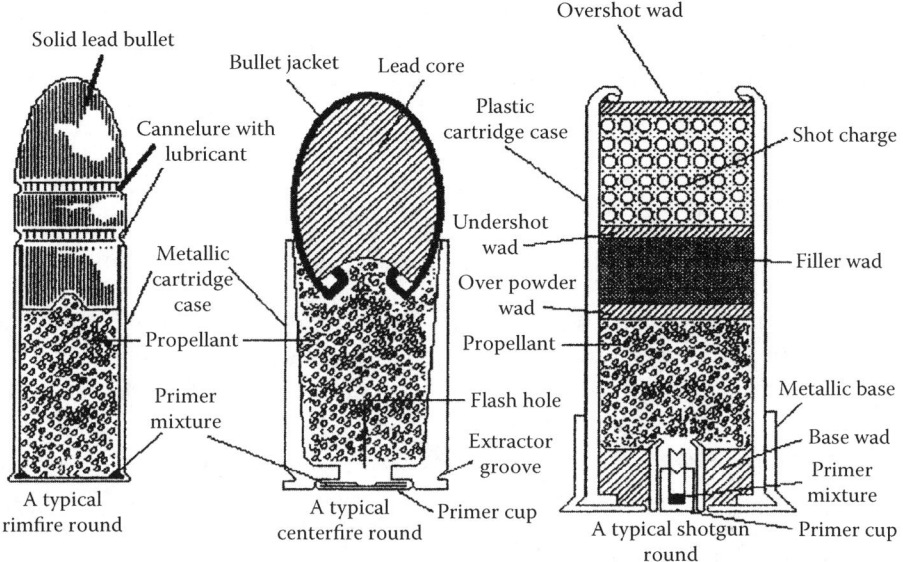

Figure 1.1 Typical ammunition types.

(c) Ammunition

The Oxford dictionary defines a cartridge as "a case containing a charge of propellant explosive for firearms or blasting, with bullet or shot if for small arms." Other terms, such as ammunition, round of ammunition, or round, are also used for cartridges and are equally acceptable. Bullet, however, is wrongly used in this context and should be reserved for the projectile only.

A round of ammunition consists of a primer, propellant, and bullet, all of which are contained by a cylinder-shaped cartridge case (shell, case). Instead of a single bullet, shotgun ammunition typically contains numerous spherical lead balls which are totally enclosed within the cartridge case. Shotgun cartridges are usually made of plastic with a metal base. Cartridges for rifled firearms are usually made of brass with the base of the bullet inserted into the neck of the cartridge case.

Figure 1.1 gives cross-sectional views of ammunition for rifled bore and smooth bore firearms.

(d) Discharge of a Firearm

The firing mechanism of a firearm consists of a mechanical device which causes a hammer to fly forward and deliver a blow to the firing pin when the trigger is pulled. In some firearms the hammer and firing pin are made in one piece. The firing pin goes through a small hole in the breech face and strikes the primer cup. [The well-known phrase "lock, stock and barrel" originates from firearms—the lock is the mechanism of the firearm, the barrel is the tube through which the bullet travels, and the stock is the means of holding the firearm (butt/handle/grip)].

The primer cup contains a mixture of chemicals which sensitize each other to percussion and rapid burning, and consequently the primer burns rapidly producing a flame and a shower of hot particles that penetrates and ignites the propellant.

The burning of the propellant very rapidly produces a large volume of gases in a confined space accompanied by a substantial temperature and pressure rise. The resultant gas pressure forces the bullet away from the cartridge case and down the barrel of the firearm. The temperature and pressure rise also serves to cause the cartridge case to expand in the chamber, thereby effectively sealing the chamber to prevent any rearward escape of gas (obturation), which would lead to a reduction in pressure and consequently a reduction in bullet velocity.

The time span from the firing pin hitting the primer cup to the bullet leaving the gun is typically in the region of 0.01 to 0.03 seconds.[4] Muzzle velocities range from approximately 600 feet per second for very low power handguns to approximately 3,500 feet per second for very powerful rifles. Temperatures and pressures inside a gun during discharge can be in the region of 3,000°C[5] and 50,000 pounds per square inch.[6]

References

1. Firearms Act (Northern Ireland) (London: HMSO, 1969), chapter 12.
2. Association of Firearm and Toolmark Examiners, *Glossary*, 3rd ed. (Chicago: Available Business Printing, 1994).
3. Major Sir Gerald Burrard, *The Modern Shotgun,* vol. 1, *The Gun* (Southampton, UK: Ashford Press, 1985), 17.
4. *Textbook of Small Arms* (London: HMSO, 1929), 267.
5. C. L. Farrar, and D. W. Leeming, *Military Ballistics—A Basic Manual* (London: Brassey's Publishers), 18.
6. Major General J. S. Hatcher, *Hatcher's Notebook,* 3rd ed. (Stackpole Books), 198.

Historical Aspects of Firearms and Ammunition

II

History of Gunpowder 2

A mixture known as black powder revolutionized the art of warfare whenever it was applied to the propulsion of missiles. Black powder is a mixture of potassium nitrate (saltpeter), charcoal, and sulfur in varying proportions, granulation, and purity. A typical composition of a modern black powder is saltpeter 75%, charcoal 15%, and sulfur 10%.[7] A mixture of saltpeter, charcoal, and sulfur with other ingredients was used in China and India in the eleventh century for incendiary and pyrotechnic purposes long before "true" black powder was invented.[8] History often deals in conjecture and opinion and it is not known for certain when and by whom black powder was invented, or when and by whom it was applied to the propulsion of a missile from a firearm.* The composition of black powder was first recorded by English Franciscan monk Roger Bacon in 1249, but he did not apply it to the propulsion of a missile from a firearm. This use of black powder is usually credited to a German Franciscan monk Berthold Schwartz in the early fourteenth century.[9]

Whenever black powder was used as a propellant in guns it was commonly referred to as gunpowder. At first the ingredients were simply mixed together, but the resulting gunpowder had a tendency to separate into its component parts when carried, and it also absorbed moisture. Also the purity of the ingredients varied markedly, and a combination of these factors led to the relative unreliability of early gunpowder.

Improved methods of combining the chemicals evolved and by the fifteenth century a form known as "corned" gunpowder had been developed in which the components were bonded together in small granules.

For many years experiments were conducted to determine the best composition of the mixture for use in firearms. Some examples of the formulas used at various times are:

---

* It is interesting to note that prior to the introduction of modern methods to determine the alcohol content (proof) of distilled spirits, black powder was used for this purpose. Equal amounts of the alcoholic drink and black powder were mixed and set on fire. If it did not burn, it was "underproof" and did not contain enough alcohol. If it burned with too bright or too yellowish a flame, it was "overproof" and contained too much alcohol. If it burned with a steady blue flame, it was correct.

	% Saltpeter	% Charcoal	% Sulfur
c. 1253, Roger Bacon	37.50	31.25	31.25
1350, Arderne	66.6	22.2	11.1
1560, Whitehorne	50.0	33.3	16.6
1560, Bruxelles studies	75.0	15.62	9.38
1645, British Government Contract	75.0	12.5	12.5
1781, Bishop Watson	75.0	15.0	10.0

Note: Other formulas are used for blasting purposes and for pyrotechnic devices.

Any marked deviation from the last two formulas produces gunpowder which has a slower burning rate or which burns with less vigorous effect.[10]

Black powder was used as a firearms propellant until it was gradually replaced by smokeless propellants toward the end of the nineteenth century.

References

7. T. L. Davis, *Chemistry of Powder and Explosives*, 3rd ed. (London: Chapman & Hall), 39.
8. W. W. Greener, *The Gun and Its Development*, 9th ed. (London: Arms and Armour Press), 13.
9. William Chipchase Dowell, *The Webley Story* (Leeds, UK: Skyrac Press), 179.
10. Davis, *Chemistry of Powder and Explosives*, 39.

History of Ignition Systems 3

Ignition of the propellant was a major problem from the introduction of gunpowder in the early fourteenth century until the development of a percussion primer by a Scottish clergyman, the Reverend Alexander John Forsyth, in 1805.[11]

The first means of igniting the propellant was by placing a glowing twig or a hot wire into a touch hole at the rear of the barrel where it came into contact with the propellant. This direct method of ignition had many disadvantages: the firer needed to be near a fire, ignition was at the mercy of the wind and the rain, and it was difficult to aim properly.

To overcome the lack of mobility the "slow" match was developed. The match consisted of a piece of cord which had been soaked in a strong solution of potassium nitrate and then dried. When placed in the touch hole and lit, the match would smolder with a glowing end at the rate of about an inch a minute until it reached and ignited the propellant.[12] Speed of ignition and dependence on weather conditions were serious disadvantages.

The first mechanical device to achieve ignition was the matchlock which derived its name from the "slow" match. The match was attached to the gun by a match holder and the action of the trigger lowered the glowing end of the match into a flash pan which contained loose gunpowder (priming powder). The powder in the pan was ignited (flashed) by the glowing match end and the flame was passed through a small barrel vent to ignite the main propellant charge. This was a major improvement in ignition systems as the time of discharge closely coincided with the pull of a trigger. Early matchlocks had an open flash pan and consequently a sudden gust of wind could remove the gunpowder from the flash pan. This was partly solved by fitting a cover over the flash pan, although the smoldering match system of ignition was still dependent on weather conditions.

The next major improvement in ignition systems was the wheellock. It worked in the same way as the matchlock by conveying the flame from the gunpowder in the flash pan through a barrel vent to ignite the main propellant charge. However, the ignition of the powder in the flash pan was achieved by sparks from flint stones or pyrites being held by a moving arm drawn down against a spring-operated, spinning serrated metal wheel.[13] The spring for the wheel had to be tensioned with a key before firing each shot.

A more reliable and important variation of the principle of the wheellock was the flintlock. In the wheellock sparks were produced by a grinding motion whereas in the flintlock system sparks were produced by a striking motion.

A cock or hammer with a piece of beveled flint securely clamped to it and an L-shaped steel flash pan cover, called a frizzen, completed the spark making battery. When the hammer fell, the flint struck the upper face of the hinged pan cover (forcing it open and exposing the gunpowder in the pan) and caused sparks which ignited the gunpowder in the pan. Again, the flame from the gunpowder in the pan was directed through a small vent in the barrel causing the main propellant charge to ignite.[14]

All the means of ignition, from the hot wire to the flintlock, were dependent to a greater or lesser extent on the weather conditions and none offered the reliability of ignition experienced with modern ammunition. However, the flintlock was a very efficient mechanism, and with the introduction of a waterproof flash pan in 1780,[15] the flintlock offered the shooter a reasonably reliable means of ignition under most weather conditions.

The flintlock was not without its faults. Misfires were not uncommon and since each piece of flint was serviceable for only 20 to 30 shots, the ignition system had to be efficiently maintained. The priming powder remained a potential weakness since wind or rain could dispose of it at the crucial moment of firing. Also the small time delay between pulling the trigger and the ignition of the main propellant charge was annoying. It took time for the flint to scrape along the frizzen and for the sparks to fall into and ignite the main propellant charge. The shooter had to make allowance for the delay especially when aiming at moving targets. A quicker and more reliable means of igniting the main propellant charge was needed.[16]

According to many writers, the Reverend Alexander Forsyth studied a group of chemical compounds called metallic fulminates whose existence had been known from 1800. It was also known that they exploded with a flash when struck a sharp blow with a hard object. In 1805, he applied this property of metallic fulminates to firearms ignition, thereby inventing the percussion system of ignition.

In 1807 he took out a patent on his invention by which a pivoted magazine deposited a few grains at a time of mercury fulminate into a touch hole in the barrel of the firearm. The mercury fulminate was detonated by a blow from the hammer of the firearm sending flame through the touch hole to ignite the propellant. "Instant" ignition had been achieved.

The pivoted magazine was too complicated and subsequent development by other workers was geared toward a more convenient and efficient means of presenting the mercury fulminate to the firearm.

This led to several short-lived innovations including the tubelock, patch-primers, and the pill-lock eventually leading to the percussion cap, which proved to be the most efficient and practical way to package the primer. The development of the percussion cap (small waterproof copper cups) is credited to Joshua Shaw in 1816.[17] The cap (primer) was placed over a permanent hollow nipple, screwed into a flash hole in the gun barrel, and detonated by the crushing impact of the hammer.

References

11. *Textbook of Small Arms* (London: HMSO, 1929), 2.
12. Frederick Wilkinson, *Firearms* (Rochester, NY: Camden House Books), 4.
13. Edsall James, *The Story of Firearms Ignition* (Curley Printing), 4.
14. Joseph G. Rosa, and Robin May, *An Illustrated History of Guns and Small Arms* (Cathay Books), 22.
15. William Chipchase Dowell, *The Webley Story* (Leeds, UK: Skyrac Press), 193.
16. Frederick Wilkinson, *Firearms* (Rochester, NY: Camden House Books), 44.
17. W. W. Greener, *The Gun and Its Development,* 9th ed. (London: Arms and Armour Press), 115.

History of Bullets

4

The first projectiles to be discharged from any type of firearm were stones, feathered iron arrows, and iron shot. These were discharged from cannons.[18] The first handheld firearms had bore diameters between 1.5" and 2.0" and suitable small round stones were used as projectiles. The earliest use of lead in bullets would appear to have been about 1340 and these consisted of spherical lead bullets.[19] Firearms of this era were large and heavy and from this time to the present day there has been a gradual reduction in bore size and weight. By the time the flintlock pistol came into use, the spherical lead bullets were between 0.6" and 0.7" in diameter.* Bullets of this type were used for many years in smooth bore muzzle-loading firearms where the bullet did not have to be a tight fit in the bore.

Rifling of the bore was found to improve the accuracy and consequently the effective range of firearms, and was first applied to firearms by German gunsmith Augustin Kutler in 1520.[20] The introduction of rifling coupled with the development of breech loading firearms focused attention on bullet design.

With rifled bore firearms the bullet had to be a tight fit; otherwise it would not grip the rifling when discharged but, if it was too tight a fit, it was difficult to load the gun. The problem of bullet size created particular loading problems for rifled muzzle-loaded firearms and for rifled breech-loading firearms. A tight-fitting spherical lead bullet was difficult to load, especially when the bore or chamber was dirty with fouling from previous shots.

The first attempts to solve the problem involved the use of a belt (driving band) around the lead ball. The bullets were cast in a mold and the lead belt was an integral part of the bullet. The spherical part was an easy fit in the bore and the belt was made large enough to fit the rifling. This bullet proved to be unsatisfactory as the belt caused the bullet to tilt after leaving the muzzle and it was very susceptible to the effects of the wind, causing poor accuracy.

During this period a large number of bullet designs were produced and tested, and it was found that an elongated bullet was much more efficient than a spherical one. The elongated bullet had greater weight for a given diameter and was more stable in flight. In 1855, General J. Jacob produced a cylindro-ogival bullet with four cast-on lugs to engage the rifling. Another

---

* The well-known term "biting the bullet" means exactly what it says and originates from surgery performed on or near the battlefield before the advent of anesthetics, the injured party bracing himself by biting on a bullet.

mechanically fitting bullet was produced by English engineer Joseph Whitworth who developed a hexagonal bored barrel and hexagon-shaped bullets. The hexagonal bullet had six flat portions along its cylindrical body which were given a twist corresponding to that of the rifling. Such mechanically fitting bullets were very difficult to manufacture and were soon found to be unnecessary. By this time the bullet diameter had been reduced to .45".

The first practical solution to the problem was developed in 1849 by Captain Minie of the French Army. He produced a cylindro-ogival bullet with a tapered hollow base containing a semispherical iron cup. When the gunpowder burned, the hot expanding gases forced the iron cup into the bullet, which spread the bullet slightly so that its sides gripped the rifling. It was soon discovered that the same effect could be obtained without the iron cup and the Minie bullet was abandoned.

In 1863, William Ellis Metford produced a cylindro-conoidal bullet with a shallow depression in its base. The bullet was made of lead hardened with antimony and the cylindrical part was wrapped in a sheath of paper. The shape and design of this bullet resembles the modern bullet.[21]

Another problem related to bullet design was the fact that the rifling could cause lead to be stripped from the bullet, resulting in "leading" of the bore, which has a detrimental effect on accuracy by deforming the bullet and reducing the efficiency of the rifling. The use of antimony or tin to harden bullet lead dates from the early nineteenth century. The use of hardened rather than soft lead serves to reduce leading of the bore and deformation of the bullet and also slightly reduces the extent of bullet deformation on hitting a target.

The use of hardened lead did not eliminate leading but it slightly reduced the extent of the problem. Lubrication of the bullet was found to significantly reduce the amount of leading by preventing the partial melting of the lead by heat due to friction. Lubricants such as tallow and beeswax were placed in annular grooves at the rear of elongated bullets.

The problems of leading and bullet deformation were eventually eliminated by the use of a bullet jacket (envelope). Such a bullet was introduced in 1883 by Major Rubin of the Swiss Army and consisted of a soft lead core covered with a copper jacket. This was an important step in bullet development because up to this time the rate of the rifling twist was limited by its effect on the unjacketed lead bullet. With this new bullet the rate of twist could be substantially increased and the rifling grooves could be made shallower. (The rate of rifling twist can be altered to give different rates of spin of the discharged bullet.)

A bullet jacket is normally harder than the bullet core material but soft enough to take up the rifling and not cause excessive wear to the barrel. Bullet jackets were for a long period made of cupronickel (80% copper, 20% nickel), gilding metal (90% to 95% copper, 10% to 5% zinc) or steel which was coated with a softer metal to prevent barrel wear and rusting.[22] In 1922, 1% to 2% of tin was added to the gilding metal because of its lubricating properties.

Unjacketed lead bullets are unsuitable for use in most modern self-loading firearms. With higher-velocity firearms, melting and fusing of the exposed lead surface can occur causing leading of the barrel, deformation of the bullet, and a loss of accuracy of the firearm. Modern lubricated unjacketed lead bullets are usually confined to use in lower-velocity revolvers and 0.22" caliber rimfire rifles and pistols, that is, firearms with a muzzle velocity of less than about 1,200 feet per second. Another important factor influencing the use of unjacketed lead bullets is that they are more prone to "feeding" problems in self-loading firearms because the exposed part of the unjacketed lead bullet is more susceptible to damage than its jacketed equivalent.

The vast majority of modern bullet types is either completely or partially jacketed, usually with gilding metal, and is produced in a range of shapes, sizes, weights, and designs depending on their intended use.

References

18. Joseph G. Rosa, and Robin May, *An Illustrated History of Guns and Small Arms*.
19. William Chipchase Dowell, *The Webley Story* (Leeds, UK: Skyrac Press), 198.
20. Dowell, *The Webley Story*, 198.
21. Dowell, *The Webley Story*, 202.
22. P. J. F. Mead , *Notes on Ballistics*, 6.

History of Ammunition 5

The self-contained metallic cartridge is a relatively recent development in historical terms. Gunpowder has been in use as a firearms propellant for about 670 years, but the metallic cartridge is only about 160 years old. The modern self-contained metallic cartridge was perfected about 122 years ago and high-velocity types with smokeless powders were developed about 102 years ago.[23]

Prior to the introduction of a self-contained cartridge, firearms were muzzle-loaded by pouring a measured amount of gunpowder down the barrel followed by the bullet and then compacting the gunpowder/bullet combination by the use of a plunger and some sort of wad. Ignition of the gunpowder was accomplished separately. Obviously, this system suffered several major disadvantages. Faster reloading in order to achieve greater firepower was desirable, the means of ignition was susceptible to weather conditions, and it was necessary to carry items of equipment ancillary to the firearm, that is, gunpowder, ignition powder (finely powdered gunpowder), bullets, wads, and ramming rod. Because of the long loading time, the advantages of a self-contained ammunition package were evident early in the history of firearms and many attempts were made to produce such a package.

One of the earliest attempts to decrease loading time was a breech-loading matchlock firearm with the rear end of the barrel counter bored to give a larger diameter than the rest of the bore. A removable iron chamber complete with its own flash pan and loaded with gunpowder and bullet was inserted. Extra loaded insert chambers could be carried.[24]

A paper cartridge was developed about 1550 and consisted of gunpowder and bullet wrapped in a cylinder-shaped paper package or a small paper bag of gunpowder attached by thread to the bullet. In use the bottom of the paper cartridge was torn open (usually with the firer's teeth), and the gunpowder and bullet poured down the barrel from the muzzle end after placing a small amount of gunpowder in the flash pan. The paper was sometimes rammed down the barrel and used as a wad to prevent the bullet dropping out of the barrel.

Various designs of the paper cartridge were in general use by the middle of the seventeenth century and paper was used for cartridge manufacture for about 300 years.

Whenever the complete cartridge, including paper, was loaded into the firearm, the paper cartridge case burned when the charge was fired. However, smoldering pieces of paper could remain in the barrel and on reloading an explosion could occur. This led to the introduction of a completely combustible paper cartridge, with the paper nitrated prior to assembly. Nitrated animal intestines were also used in cartridges during this period. The paper cartridges caused problems in damp weather and the cartridges had to be carried in waterproof containers. Several attempts were made to waterproof paper cartridges using varnish but this did not achieve widespread acceptance.

The earliest example of a fully self-contained cartridge was produced by Swiss engineer Jean Samuel Pauly in 1808. This cartridge was loaded directly into the breech of a firearm, which was also developed by Pauly, and was fired by a needle piercing it.

An improved form of the cartridge was patented by Pauly in 1812. It consisted of a paper body rolled around the front portion of a rimmed brass base piece, the base of which had a central recess to contain the primer powder, which was sealed with a small piece of gummed paper to retain it in position and protect it from moisture.[25]

This was one of the most important developments in firearm history and is the earliest example of a fully self-contained centerfired cartridge. However, the system did not gain widespread acceptance as it applied only to firearms of Pauly's design. It did establish the principle of a completely self-contained cartridge, that is, a cartridge with its own means of ignition as an integral part. Figure 5.1 gives a cross-sectional view of the Pauly cartridge.

The next cartridge with an integral primer was the needle-gun cartridge developed by a Prussian, Johann Nikolas Dreyse, in 1831. In its original form it was made with either a paper or linen envelope and in its later form it was made entirely of paper. It had a flat base and was tied shut above the bullet which was contained in a sabot. There was a recess in the base of the

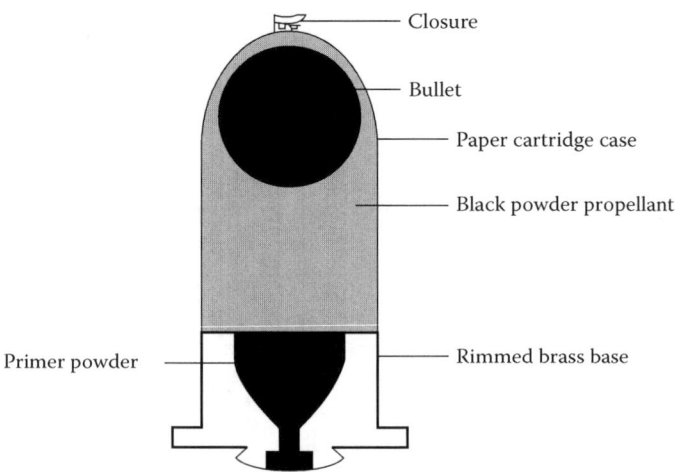

Figure 5.1 Pauly cartridge.

History of Ammunition

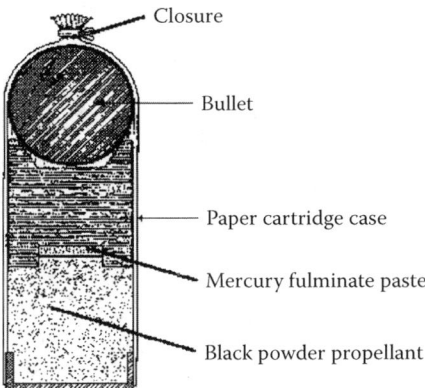

Figure 5.2 Dreyse cartridge.

sabot that contained a mercury fulminate paste. Ignition of the cartridge was accomplished by a long spring-operated needle which had to penetrate the full length of the gunpowder charge to reach the mercury fulminate. Figure 5.2 gives a cross-sectional view of the Dreyse cartridge.

Further development of the needle-gun concept led to a pasteboard cartridge case with the primer in the base portion in the form of a shallow metal foil cup containing a flanged percussion cap with its open end facing the base of the cartridge. A small hole was made through the center of the base and metal foil cup, and the cap was ignited by the penetration of a short firing needle.

An innovation not involving a conventional cartridge case was introduced by Joseph Rock Cooper in 1840. This was a bullet with a charge of gunpowder placed in a cavity at the base of the bullet. Ignition was by means of an external percussion source. Development of this concept by other workers culminated in a cone-shaped bullet with a charge of gunpowder in a base cavity which was closed by a cork plug and fitted with a priming system.[26]

As there was nothing to prevent the rearward escape of gas and as the bullet itself had only about 1/15 its weight in gunpowder charge, the bullet lacked power. Misfires were common and the system was abandoned about 1856.

Until 1846 all attempts to develop a satisfactory fully self-contained cartridge shared a serious disadvantage. None of them effectively sealed the chamber at the time of discharge and consequently there was a rearward escape of gas resulting in a reduction in the efficiency of the system. This problem was solved by the introduction of the metallic cartridge case which momentarily expands during the discharge process and seals the chamber.

The first recorded examples of fully self-contained completely metallic cartridges were the pinfire cartridges of the early 1850s. These consisted of a thin copper cartridge case with a striker pin projecting radially from the base end (brass came into general use in the 1870s and replaced copper as the case

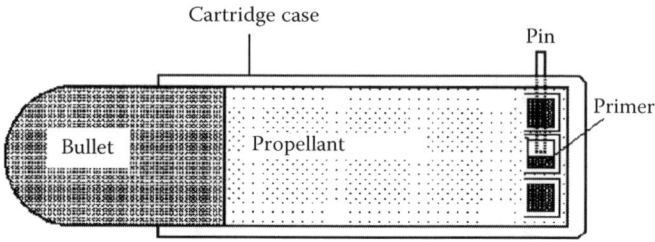

Figure 5.3 Pinfire cartridge.

material). The striker pin was aimed at the priming composition but positioned just clear of it. Figure 5.3 gives a cross-sectional view of the pinfire cartridge.

By effectively sealing the bore during discharge the pinfire cartridges made breech loading a much more practical proposition and these cartridges were manufactured until the late 1930s. A major disadvantage of this system was that because of the projecting pin the cartridge could be loaded in one position only.[27]

The next stage in cartridge evolution was the rimfire cartridge. The idea of a cartridge with a hollow rim to contain the priming composition was patented by French gunsmith Houllier in 1846 and developed by another French gunsmith, Flobert. The cartridge was originally produced with no gunpowder charge, the priming composition serving as both igniter and propellant.

In 1854 the American firm of Smith & Wesson developed the design by lengthening the case so that it could hold a charge of gunpowder. The rimfire system became very popular and was manufactured in a range of calibers. A major advantage of the rimfire cartridge was that it made possible the construction of firearms having a supply of cartridges housed in a magazine.

As firearms developed the trend was toward smaller and greater power and range and it was found that the thin metal base of the rimfire cartridge could not withstand the higher pressures involved. This was a disadvantage that could not be readily overcome and was one of the main reasons for the decline in popularity of the rimfire cartridge. Other disadvantages of the rimfire cartridge are unsuitability of design for modern firearm loading and ejection systems, the larger amount of priming composition that is required, and the manufacturing inconvenience of ensuring an even spread of priming composition around the rim.

Rimfire cartridges are still manufactured but only in 0.22" and 0.17" calibers and all other modern firearms ammunition is centerfire (central-fire).

Centerfire cartridges were produced by Pauly in 1808 but it was not until 1854 that the firm of Smith & Wesson perfected and patented both the centerfire and rimfire metallic cartridge case. Since this time cartridge development has consisted of many small improvements, some resulting from advances in engineering and metallurgy, some from improvements in firearms design, and some as a result of the development of modern smokeless propellants.

The modern cartridge evolved over this period to a very high standard of reliability. Ironically, serious attempts by reputable large munitions companies are now being made to perfect a completely combustible cartridge and/or caseless ammunition which would be suitable for use in modern firearms.[28]

References

23. Frank C. Barnes, *Cartridges of the World*, 3rd ed. (Digest Books), 3.
24. William Chipchase Dowell, *The Webley Story* (Leeds, UK: Skyrac Press), 204.
25. W. H. B. Smith, and Joseph E. Smith, *The Book of Rifles* (Castle Books), 29.
26. Dowell, *The Webley Story*, 218.
27. Dowell, *The Webley Story*, 219.
28. Ivan V. Hogg, *Encyclopedia of Modern Small Arms* (London: Hamlyn/Bison), 77.

History of Firearms 6

The history of firearms is long and complicated, encompassing innovations and developments in ammunition: from crude black powder muzzle loaders to modern brass-cased, centerfire cartridges using smokeless propellant, in ignition systems; from a glowing twig touching gunpowder through a simple flash hole in the barrel to a firing pin which strikes and crushes the priming mixture thereby "instantly" igniting the propellant charge, in mechanical developments; from the simple metal tube attached to a stick to the finely machined high technology firearms which are capable of operating from single shot to fully automatic fire, in metallurgy; from crude iron, which withstood the pressure of the weak early powders, to high tensile metals that can withstand pressures in the tens of thousands of pounds per square inch. The history of ammunition closely parallels the history of firearms as one is designed to accommodate the other.

Firearms were in general use in Europe for two centuries before the introduction of printing; consequently reliable accounts of early arms development are rare. The first firearms were cannons that fired large round stones, iron balls, or a quantity of arrows and would appear to have been introduced into Europe from the Eastern nations around 1300. Early cannons were small, and shot arrows weighing about half a pound, although very large cannons weighing about 4 tons and firing stone shot weighing in the region of 350 pounds were also produced.

The first handguns were really handheld cannons now known as cannonlocks, with the lock the means of firing the gun. These obviously evolved from the early small cannon and were first used in Europe about 1324.[29] The cannonlock had a cylindrical metal barrel, about 9 to 12 inches long, attached to a staff or pike. They were muzzle-loaded with gunpowder; wad; and round stones, metal balls, or bolts (similar to crossbow bolts). In use, they were crudely aimed with one hand, with the staff held under the arm and fired by a glowing twig or hot wire brought into contact with the gunpowder through a touch hole in the barrel.

Hand cannons developed through various stages. They were shortened and redesigned for use from horseback and were used in combination weapons where the weapon could either be used as a firearm or, for example, as a club or an axe. Many different designs of hand cannon were widely used for many years, until the middle of the fifteenth century when they were completely superseded by the matchlock.

The matchlock was developed about 1400,[30] and by mechanically carrying the fire to the priming mixture it made possible the fitting of elementary aiming sights to the firearm. Early handguns consisted of a barrel secured to a wooden or metal arm, but with the introduction of the matchlock musket, firearms became much more sophisticated and began to resemble the modern rifle. Numerous variations of the matchlock were produced and were used for many years, until it was eventually superseded in the seventeenth century by the wheellock and the flintlock.

The wheellock was developed about 1515.[31] This was an important development in firearms as, apart from dispensing with the need for a glowing match, the wheellock mechanism could be produced in any desired size which made possible the production of pistols small enough to be carried about the person. As with the hand cannon, combined wheellock weapons were produced where pistols were attached to weapons such as maces, swords, and crossbows. The wheellock mechanism was intricate and subject to mechanical failures which were difficult to repair. This prompted a search for a simpler, more reliable mechanism, resulting in the introduction of the flintlock.

The flintlock was developed about 1525[32] and used a simpler and much more reliable mechanism than the wheellock. The flintlock was used successfully until it was generally superseded by the percussionlock (caplock mechanism) about the middle of the nineteenth century. A measure of the success of the flintlock is demonstrated by the fact that, until 1935, they were made in Germany and Belgium for export to Africa and Asia.[33]

The percussionlock was developed in 1805, and by 1816 had evolved into a simple and reliable form. The percussionlock was the predecessor of the modern firearm and used a priming cap consisting of a small metal cup in the base of which was a dried paste containing mercury fulminate. This was placed over a permanent hollow nipple leading to the gunpowder so that the mercury fulminate paste would be crushed between the base of the cup and the nipple by the striking action of the hammer of the firearm. This produced a flame that passed through the hollow nipple and ignited the gunpowder.

The modern firearm employs the percussion principle but the percussion cap (primer) is an integral part of the round of ammunition.

The first practical repeating firearm was a revolver manufactured by Samuel Colt in 1835.[34] Up to this time the vast majority of firearms were single shot. This was a serious disadvantage as the firer was defenseless for a period of time while reloading. However, the introduction of this revolver heralded the first practical multishot firearm. The revolver principle was not new, as flintlock revolvers were produced prior to 1650.[35] However, these were not practical firearms as they were very prone to mechanical failure.

When a bullet leaves the muzzle of a firearm there is recoil in the opposite direction to the travel of the bullet. Although the recoil is a nuisance it can be used to eject the spent cartridge case, load a live round of ammunition, and cock the mechanism. This can also be achieved by using some of the gas generated during discharge.

In the self-loading system the block or slide that moves backward and forward is stopped after each cycle and stays stopped until the trigger is pulled again. This mechanism can be modified so that the firearm continues to fire until either the ammunition is expended or the trigger released (automatic fire). Some firearms incorporate a selector lever, which allows them to deliver either a single shot or a burst of a preset number of shots, or to become fully automatic.

As early as 1718 there was a hand-operated repeating gun, and in 1862 Dr. Richard Gatling demonstrated a weapon of this kind which used revolving barrels. These weapons had severe limitations and it was not until 1884 that the first real fully automatic machine gun was patented by Sir Hiram Maxim. This was the first automatic firearm, and it was recoil operated.[36] The development of the Maxim machine gun focused attention on the development of self-loading rifles and pistols.

The rifle evolved from the musket which was a long-barreled firearm with a fore end or forearm extending nearly to the muzzle. Dozens of designs of self-loading rifles were produced. One of the first practical designs was developed in Austria by Mannlicher in 1885 and it was recoil operated.[37]

The recoil-operated self-loading system was incorporated in the first successful multishot pistol which was designed by Hugo Borcharott, and marketed in 1893. George Luger modified the design and produced a highly successful pistol which was in production until 1942.[38]

Today, self-loading firearms are either recoil operated or gas operated, and progress since the production of the Maxim machine gun has consisted mainly of a series of mechanical improvements resulting in the modern, highly reliable, self-loading, semiautomatic or fully automatic firearms now employed throughout the world.

References

29. W. H. B. Smith, and Joseph E. Smith, *The Book of Rifles* (Castle Books), 8.
30. Frederick Wilkinson, *Firearms* (Rochester, NY: Camden House Books), 4.
31. Smith and Smith, *The Book of Rifles*, 18.
32. Smith and Smith, *The Book of Rifles*, 21.
33. Smith and Smith, *The Book of Rifles*, 25.
34. Joseph E. Smith, *Book of Pistols and Revolvers* (Castle Books), 20.
35. Smith, *Book of Pistols and Revolvers*, 19.
36. Major General J. S. Hatcher, *Hatcher's Notebook,* 3rd ed. (Stackpole Books), 32.

37. W. W. Greener, *The Gun and Its Development*, 9th ed. (London: Arms and Armour Press), 731.
38. Smith, *Book of Pistols and Revolvers*, 28.

Chemical Aspects of Firearms and Ammunition

III

Cartridge Cases   7

The cartridge case is designed to house the primer, propellant, and to securely retain the bullet in the neck of the case. Cartridge case design is affected by various factors, the most important being:

The role of the ammunition.
Type of weapon.
Design of the bullet to be used.
Type of ignition system, that is, Boxer primed or Berdan primed.

The vast majority of cartridge cases are made of brass (approximately 70% copper and 30% zinc) but other materials such as steel, coated with either zinc, brass, gilding metal, copper, lacquer or blackened; copper; nickel-plated brass; cupronickel (approximately 80% copper and 20% nickel); gilding metal (approximately 90% copper and 10% zinc); aluminum. Teflon-coated aluminum and plastic are also encountered.

Second in popularity to brass is steel. One specification for cartridge case steel is carbon 0.08% to 0.12%, copper 0.25%, manganese 0.6%, phosphorus 0.035%, sulfur 0.03%, and silicon 0.12%.[39]

Shotgun cartridges are usually plastic with a brass or coated steel base, but paper with a brass or coated steel base and all plastic shotgun cartridges are also encountered. "All brass" shotgun cartridges are also known in older ammunition, and are currently manufactured for reloading purposes. Some .410" caliber shotgun cartridges are all aluminum.

Apart from shotgun cartridges, brass is by far the most common material used for the manufacture of cartridge cases. Experience has proved brass to be the most suitable as it is strong, sufficiently ductile, nonrusting, suited to drawing operations during manufacture, of reasonable weight, and readily available.

The strict specifications and quality control procedures for cartridge manufacture reflect the very important role the cartridge case plays in the discharge process. In addition to housing all the components of a round of ammunition in one package, a cartridge must:

Be safe to store, transport, and use.
Seal against ingress of moisture and oil.

Consistently achieve the required ballistics performance even under very different climatic conditions.

Maintain performance after many years of storage.

Be sturdy enough to withstand rough treatment, especially on the battlefield.

Achieve moderate and consistent chamber pressures.

Function satisfactorily and reliably from belt- and magazine-fed firearms under sustained fire conditions.

Be relatively cheap and readily manufactured during periods of emergency such as war.

Retain the bullet for a period after ignition to allow the propelling gas pressure to build up sufficiently to achieve peak performance.

Effectively seal the chamber during discharge.

The type of brass used is very important to the manufacturing process and manufacturers carefully specify the quality of the brass to be used. Four examples of specifications are as follows[40]:

1. 68% to 74% copper and 32% to 26% zinc. Impurities must not exceed 0.2% nickel, 0.15% iron, 0.1% lead, 0.05% arsenic, 0.05% cadmium, 0.008% bismuth. Tin and antimony must be absent and there must not be more than a trace of any other impurity.
2. 65% to 68% copper and 35% to 32% zinc with up to 0.2% nickel. There must be no individual impurity in excess of 0.1% and no more than 0.1% lead, 0.05% iron, and 0.03% phosphorus.
3. 70% copper and 30% zinc with not more than 0.25% of all other impurities combined.
4. 72% to 74% copper and 28% to 26% zinc. Impurities must not exceed 0.1%, and there must not be more than 0.1% lead and 0.05% iron.

When a round of ammunition is discharged in a firearm, the internal gas pressure, and to a much lesser extent the temperature rise, causes the cartridge case to expand tightly against the chamber walls (obturation). This is an extremely important function of the cartridge case as this prevents the rearward escape of gas. Such an escape of gas would reduce the velocity of the projectile and consequently the efficiency of the firearm and could possibly cause a malfunction in the firearm mechanism.

If the brass in the cartridge case is too soft, it will not spring back from the chamber walls, which will probably make extraction of the spent cartridge case very difficult. If the brass is too hard, the cartridge case could crack because it is too brittle and jam the firearm mechanism. When the brass in the cartridge case is of the correct hardness, it springs back to its near original dimensions and the spent cartridge case is readily extracted.

For higher-velocity ammunition the hardness of the brass usually decreases from the base to the neck of the cartridge case. Cartridge cases for low-velocity ammunition are normally made to a standard hardness along their entire length.

The base of a cartridge case must be strong enough to withstand ramming and extraction (this can happen numerous times to an individual round of ammunition during loading and unloading procedures) while the neck of the case must be strong enough to rigidly support the bullet yet flexible enough to expand and seal the chamber during discharge. High-temperature discharge gases can raise the pressure inside the cartridge case to 40,000 pounds per square inch in a very short time period.[41]

As the cartridge case is subjected to considerable stresses during loading, firing, and extraction, case thickness as well as case hardness needs to be carefully controlled. There must be a sufficient thickness of metal at the base of a cartridge to sustain the severe back thrust that occurs during discharge. If the metal walls of the cartridge case are too thin, the extension of the cartridge case due to longitudinal stress may cause it to fracture or the wall to separate from the base. For these reasons the thickness of metal in a cartridge case is carefully controlled and decreases from the base to the neck. The need to keep the weight to a minimum is another factor that is taken into account at the design stage.

A large quantity of propellant is required for modern high-velocity ammunition and this is accommodated by enlarging the diameter of the case over most of its length before markedly reducing the diameter at the forward end to accept the bullet. High-velocity cartridge cases are tapered and necked to avoid extraction difficulties which would be experienced if cylindrical cases were used in firearms with high chamber pressures. Most low-velocity cartridge cases are also slightly tapered.

The feeding and extraction mechanism of the firearm coupled with the type of ignition system dictates the design of the base of the cartridge case. Nearly all cartridge cases have the outside surface of the base indent-stamped by the maker (head stamp). Information such as the maker's initials, code, or mark, year of manufacture (mainly military ammunition), caliber or other coded information are indent-stamped into the base. It is sometimes possible, even for old ammunition, for a manufacturer to check its records and give the complete specification of a round of ammunition from the head stamp details.

The joint between the primer cup and the outside of the cartridge case base is sealed with lacquer to prevent the ingress of moisture and oil. The lacquer is sometimes color coded as an aid to visual inspection during manufacture, and also sometimes to identify the type of bullet, for example, ball, tracer, armor piercing. Sometimes the mouth of the case is internally varnished, just before inserting the bullet, to waterproof the joint and to provide resistance to the pressure of the propellant gases. (Hornby has developed a special black nickel plating for use on all metallic cartridge cases. It is claimed to give a smoother surface than conventional ammunition leading to greater reliability with fewer weapon malfunctions.)

Bogus head stamps are sometimes encountered when a government, for political or economic reasons, is supporting a rebel cause in another country by supplying the rebels with ammunition. For obvious reasons the source of supply is disguised. This can be done by omitting the head stamp or by using fake head stamps.[42] It is not unknown for such ammunition to be head-stamped in such a way as to attempt to place the blame for supply on some other government.

References

39. Private communication, 1974.
40. Private communication, 1976.
41. S. Basu, "Formation of Gunshot Residues," *Journal of Forensic Sciences* 27, no. 1 (1982): 72.
42. P. Labbett, "Clandestine Headstamps," *Guns Review Magazine* (February 1987): 128.

Primer Cups (Caps) 8

Priming compositions for centerfire ammunition are housed in small metal cups which fit into a recess, called the primer pocket, in the center of the base of the cartridge case. In rimfire ammunition the priming composition is housed inside the cartridge case in the hollow perimeter of the base.

The ideal primer cup metal should expand easily to provide a gas-tight seal, be strong enough to withstand the blow from the firing pin, and also strong enough to withstand the "explosion" of the priming composition and the discharge gas pressure, even though it has been severely dented by the firing pin.

Primer cups are usually made of cartridge brass, although copper, nickel-plated copper or brass, copper alloy, cupronickel, and zinc-coated steel cups are also encountered. Primer cups for use with black powder were usually made of soft copper because of the weaker firing pin blow experienced with old black powder firearms, and the much lower pressures generated by black powder discharge. On the other hand, smokeless powders typically give much higher pressures than black powder and are much harder to ignite. Smokeless powders require a much "hotter" primer, which needs a much stronger blow from the firing pin. Consequently soft copper cups are suitable only for use with low-pressure ammunition.

Two specifications for primer cup metal are (a) 95% to 98% copper and 5% to 2% zinc with not more than 0.05% lead, 0.1% arsenic, 0.002% bismuth, 0.01% antimony, and no more than a trace of any other impurity[43]; (b) 72% to 74% copper and 28% to 26% zinc with the total impurities not exceeding 0.1% and not more than 0.1% lead and 0.05% iron.[44]

There are two types of primers used in centerfire ammunition which differ only in physical design. In European countries, the Berdan primer design is preferred, whereas in Canada and the United States the Boxer primer design is favored. The only difference between the two types is the design: the Berdan primer does not have an integral anvil, as the anvil is part of the cartridge case, whereas the Boxer primer has its own anvil which is inserted into the primer cup. Boxer primers are preferred because they are replaceable. Figure 8.1 shows Berdan and Boxer primers.

The Berdan cup is varnished internally when empty and after filling it is covered with a paper disc and then sealed with varnish.

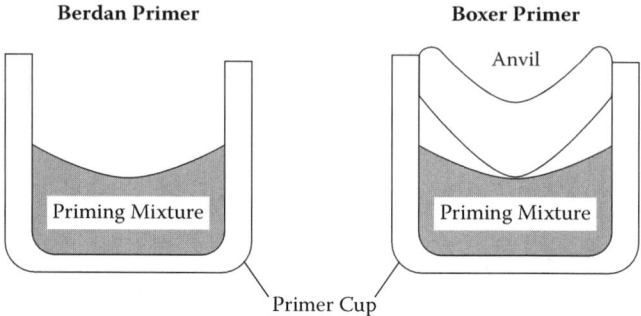

Figure 8.1 Berdan and Boxer primers.

Cupronickel and copper alloy cups that are filled with a mercury fulminate–based primer composition are closed with a tinfoil disc that is varnished on the side that is in contact with the primer composition. A varnish that is frequently used for this purpose is shellac grade 1. After fitting, the cup annulus is coated with a clear varnish to prevent the ingress of moisture or oil.

Generally speaking primer cups for rifles differ in size, structure, and amount of priming composition from those used for pistols and revolvers. Primer cups for use in rifles, pistols, and revolvers range in size from 0.175 to 0.210 inches in diameter. For shotgun cartridges, the primer cup is typically in the range 0.240 to 0.245 inches in diameter.

Although pistol and revolver primer cups may have the same diameter as rifle primer cups, rifle primer cups have a greater cup metal thickness and contain larger amounts of priming composition which is accommodated by a longer primer cup length. The increased thickness of rifle primer cups is necessary because of the heavier blow they receive from the firing pin and the higher working pressures experienced. The larger amount of priming composition is necessary because of the larger amount of propellant used in rifle ammunition.

The weight of primer composition for pistol, revolver, rifle, and shotgun ammunition can range from as little as 0.013 g to as much as 0.352 g depending on the caliber and type of ammunition, but is typically in the region of 0.05 to 0.12 g.

References

43. William Chipchase Dowell, *The Webley Story* (Skyrac Press), 267.
44. Jim Stonley, "Primers and Complete Rounds," *Guns Review Magazine* (March 1986): 166.

Priming Compositions 9

Priming compositions for firearms ammunition are mixtures which, when subjected to percussion, provide a sudden burst of flame that serves to ignite the propellant within the cartridge case. A priming composition must deliver a relatively large volume of hot gases and hot solid particles without the development of a detonating wave.

The ideal priming composition would consist of a cheap, readily available, relatively safe to handle, simple chemical compound of uniform granulation which when subjected to impact would undergo rapid, highly exothermic decomposition. The only compound to even approach these specifications is lead dinitroresorcinate; however, it is far too sensitive. In practice, no single chemical compound meets all the requirements of an ideal primer.

The next most desirable type of priming composition would be a mixture of compounds that, although individually nonexplosive, sensitize each other to ignition and rapid burning. In fact, most priming compositions consist of mixtures of one or more initial detonating agents, with oxidizing agents, fuels, sensitizers, and binding agents. The net effect of the additions is to dilute the initial detonating agent so as to convert its decomposition from detonation into rapid combustion. In some cases a single addition may serve two purposes, for example, antimony sulfide acts as a fuel as well as a sensitizer to friction, and gum arabic acts as a fuel and a binding agent. The oxidizing agents provide oxygen to support combustion of the fuel within the small space of the cartridge case. Fuels are necessary to prolong the combustion long enough to ignite the propellant. The additions may also serve to increase the volume of gases produced per unit weight of priming composition, to prevent the gases from having too high a temperature, and to contribute incandescent solid particles to the decomposition products.

The sensitivity of priming compositions varies, but that of an individual composition can also be varied to some extent by careful control of the granulation of each of the ingredients. Sometimes this is more important than the proportions of the ingredients. Nonuniformity of composition due to physical separation caused by shaking can lead to great variations in sensitivity and even failure to function. The presence of a binding agent prevents such separation as well as fixing the composition in the desired position in the assembly.

The rate of burning, volume of gases, weight of solid particles produced, and the duration of the flame are the major influences on the efficient functioning of a priming composition. For a typical priming composition of 0.15 g, the volume of gases at room temperature and pressure is in the order of 1.5 cm. The percentage of the weight carried as incandescent particles by the hot gases will vary with the composition, but can be in the region of 70%. The incandescent particles are thought to promote ignition by thermal radiation. Flame bursts from various primers were found to have effective durations varying from 400 to 750 microseconds and total durations varying from 650 to 1,500 microseconds.[45]

Generally speaking small arms primers consist of an explosive, an oxidizer, a fuel, and a frictionator. Other compounds act as sensitizers and binders.

Explosives used include azides, fulminates, diazo compounds, nitro or nitroso compounds, for example, lead or silver azide, mercury fulminate, lead styphnate, TNT, and PETN (which also act as sensitizers).

Oxidizers used include barium nitrate, potassium chlorate, lead dioxide, and lead nitrate.

Fuels used include antimony sulfide (which also acts as a frictionator), gum arabic (which also acts as a binding agent), calcium silicide (which also acts as a frictionator), nitrocellulose, carbon black, lead thiocyanate, and powdered metals such as aluminum, magnesium, zirconium, or their alloys.

Frictionators used include ground glass and aluminum powder (which also acts as a fuel).

Sensitizers used include tetracene, TNT, and PETN.

Binders used include gum arabic, gum tragacanth, glue, dextrin, sodium alginate, rubber cement, and karaya gum.

The quantity of oxidizer in the mixture is calculated to supply at least enough oxygen for the complete combustion of the primer; otherwise combustion products that are harmful to the firearm could be formed. (The frictionators could be regarded as sensitizers as they sensitize the mixture to percussion.) There may be more than one explosive, oxidizer, fuel, and frictionator in a single priming composition and sometimes a dye is added as an identifying feature or as an aid in production. Sometimes no single primary explosive is present, the mixture itself being the primary explosive.

In 1805 the Reverend Alexander Forsyth used mercury fulminate as the basis of his primer composition, and from this time the percussion system developed into today's highly reliable, universally used, percussion primer compositions. This development which started in 1805 still continues today, and manufacturers are very reluctant to release details of their compositions.

Consequently, information on primer compositions and the chemical composition of ammunition is both sparse and fragmented in the literature.

It is accepted by most writers that Reverend Forsyth's percussion priming composition was based on mercury fulminate. However, there is some respected opinion which suggests his composition was made up of wax-coated pellets of potassium chlorate mixed with combustible materials, and that it was not until 1831 that mercury fulminate was widely used as the explosive ingredient in primer compositions.[46,47]

Early priming compositions consisted of mercury fulminate and potassium chlorate along with other ingredients. With the introduction of metallic cartridge cases about 1850, it was found that brass cartridge cases were unsuitable for use with priming compositions containing mercury fulminate as the brass was embrittled due to mercury amalgamation of the zinc. This made the spent cartridge case useless for reloading purposes, and reloading was essential for economic reasons. Initially the use of copper cartridge cases solved this problem. In 1869, Hobbs, by the use of internal varnishing of brass primer cups and brass cartridge cases, made the use of brass and mercury fulminate possible by preventing the direct contact of the brass surface with the primer mix.

Whenever black powder was used as a propellant, a large amount of fouling was deposited on the inside of the barrel. On combustion, black powder produces 44% of its original weight as hot gases and 56% as solid residues in the form of dense white smoke.[48] When smokeless powders were introduced between approximately 1870 and 1890, another major problem was encountered. Smokeless powders were harder to ignite than black powder; consequently, larger priming loads were necessary for smokeless powders. Higher pressures were experienced with smokeless powders, and smokeless powders on combustion produced much less fouling than black powder. The relatively clean surfaces remaining in the barrel interior after the combustion of smokeless powder became rusted, even when the gun was cleaned immediately after use.

The cause of the rusting was traced to the potassium chlorate used in the priming composition. Potassium chloride, formed after the combustion of potassium chlorate, was deposited inside the barrel; it then attracted atmospheric moisture and caused rapid rusting of the barrel interior. Gun cleaning mixtures were organic in nature and did not dissolve the potassium chloride; consequently, despite cleaning immediately after use, salt particles trapped in the rifling and surface imperfections of the metal still caused rusting. Water proved to be efficient at removing all traces of the salt; however, it was then necessary, and very difficult, to ensure that all the water was removed from the gun; otherwise the water itself would cause rusting. The heavy residue left after the combustion of black powder substantially protected the metal surfaces from the effects of the salt, and to some extent from the effects of metallic mercury released after combustion of the primer.

The problems associated with the use of mercury fulminate and potassium chlorate led to a search for suitable alternatives, and the chemical reactions occurring within the cartridge case and the firearm were intensively studied. The objective of the study was to produce a satisfactory priming composition which was both noncorrosive and nonmercuric (NCNM).

As a result of the need to reuse spent cartridge cases for economic reasons, there has been no mercury in U.S. military small arms primers manufactured since 1898. It was used to a later date (about 1930) in certain U.S. commercial primers. In 1898 the U.S. military adopted a nonmercuric primer composition, coded H-48, for use in the .30 Krag cartridge. The primer composition was:

Potassium chlorate	49.6%
Antimony sulfide	25.1%
Sulfur	8.7%
Glass powder	16.6%

During World War I the nonmercuric primer mixture used was:

Frankford Arsenal FH-42 (1910)
Potassium chlorate	47.20%
Antimony sulfide	30.83%
Sulfur	21.97%

It was discovered in 1911 that thiocyanate–chlorate mixtures were sensitive to impact, and this led to the Winchester Repeating Arms Company's 35-NF primer composition:

Potassium chlorate	53%
Antimony sulfide	17%
Lead thiocyanate	25%
TNT	5%

After a batch of damp sulfur and/or impure potassium chlorate (polluted with potassium bromate) caused "dead" primers in millions of rounds of ammunition with Frankford Arsenals FH-42 primer mix, this primer composition was abandoned. Frankford Arsenal adopted the Winchester Repeating Arms Company's 35-NF primer mix which was then standardized as FA-70 and was used in 0.45 ACP and .30-06 ammunition throughout World War II and into the 1950s.

Priming Compositions

At this time a typical .22" caliber rimfire primer composition was the United States Cartridge Company's "NRA" which was:

Potassium chlorate	41.43%
Antimony sulfide	9.53%
Copper thiocyanate	4.70%
Ground glass	44.23%

It would appear that the Germans were approximately 23 years ahead of the Americans in the production of noncorrosive primers, despite the fact that the German compositions were published in the open literature. This may have been due to patent rights.

The first noncorrosive primer was produced by the German firm of Rheinische-Westphalische Sprengstoff AG (RWS) in 1891:

Mercury fulminate	39%
Barium nitrate	41%
Antimony sulfide	9%
Picric acid	5%
Ground glass	6%

(Barium nitrate replaced potassium chlorate.)

In 1910 the same firm produced the following .22" caliber rimfire priming composition:

Mercury fulminate	55%
Antimony sulfide	11%
Barium peroxide	27%
TNT	7%

The Swiss Army had also been using a noncorrosive primer mix since 1911:

Mercury fulminate	40%
Barium peroxide	25%
Antimony sulfide	25%
Barium carbonate	6%
Ground glass	4%

It was not until 1927 that the first American commercial noncorrosive primers appeared on the market. Some of these are as follows[49]:

	Remington Kleanbore	Western	Winchester Staynless	Peters Rustless
Mercury fulminate %	44.50	40.79	41.06	38.68
Barium nitrate %	30.54	22.23	26.03	9.95
Lead thiocyanate %	4.20	8.22	5.18	—
Ground glass %	20.66	28.43	26.66	24.90
Lead compound (?) %	—	—	—	25.91
Binder gum %	0.20	0.33	0.58	0.56

Up to this time primers had fallen into three categories: mercuric and corrosive, nonmercuric but corrosive, and mercuric but noncorrosive. Because of the disadvantages of mercury fulminate and potassium chlorate the main objective of primer development was to produce a primer with satisfactory ignition properties without the use of these two compounds. An early NCNM priming composition used copper ammonium nitrate to replace mercury fulminate, and potassium nitrate to replace potassium chlorate. The composition was:

Copper ammonium nitrate	30% to 40%
Potassium nitrate	42% to 25%
Sulfur	10% to 7%
Aluminum	18% to 28%

The first practical NCNM primer mixture with satisfactory ignition properties and good shelf life was produced by RWS in 1928. This type of primer was given the general name of "Sinoxyd" (Sinoxide/Sinoxid) and has the following general composition:

Lead styphnate	25% to 55%
Barium nitrate	24% to 25%
Antimony sulfide	0% to 10%
Lead dioxide	5% to 10%
Tetracene	0.5% to 5%
Calcium silicide	3% to 15%
Glass powder	0% to 5%

(Lead styphnate replaced mercury fulminate.)

This was the forerunner of all modern NCNM priming compositions. With very few exceptions, U.S. commercial primers became noncorrosive about 1931 but because of stringent U.S. government specifications for military ammunitions, which could not be met by the earlier versions of the new

NCNM primer mixtures, it was not until the early 1950s that U.S. military ammunition became noncorrosive. This was because early NCNM commercial priming mixtures suffered erratic ignition and unsatisfactory storage stability, and as large quantities of small arms ammunition are stored as a war reserve, military ammunition must have unquestioned reliability and storage stability.

In the United Kingdom both commercial and military ammunition used primers that were both mercuric and corrosive, until the gradual changeover to NCNM primers which was completed during the mid-1950s and early 1960s.

The explosive ingredient in Sinoxyd-type primers is lead styphnate (lead trinitroresorcinate), which is very sensitive to static electricity, and fatalities have resulted from handling the dry salt. Preparation of the pure salt is difficult, and many patented preparations, including basic modifications, exist. Some claim special crystalline forms and/or reduced static electricity hazard. Explosive ingredient substitutes for lead styphnate were sought that would be easier to make and safer to use. These included lead azide, diazonitrophenol, lead salts of many organic compounds, complex hypophosphite salts, picrate-clathrate inclusion compounds, and pyrophoric metal alloys.

In 1935 lead azide was patented for use in priming mixtures in the following mix:

Lead azide	12%
Barium nitrate	23%
Antimony sulfide	20%
Calcium silicide	10%
Tetracene	3%
Lead dioxide	20%
Lead thiocyanate	12%

In 1939 a primer mixture was patented that was identical to Sinoxyd except that diazonitrophenol was substituted for lead styphnate. Heat, humidity, and copper have a detrimental effect on diazonitrophenol and it is no longer used in primer mixes.

Normal lead styphnate has one lead atom per formula unit, whereas the basic form has two. A priming mixture using basic lead styphnate was patented in 1949:

Basic lead styphnate	40%
Barium nitrate	42%
Antimony sulfide	11%
Nitrocellulose	6%
Tetracene	1%

Other substitutes for lead styphnate included lead salts of many organic compounds, none of which gained widespread acceptance.

It was not until 1954 that preparation of the pure compound, normal lead styphnate hydrate, was accomplished. Up to this time the impure salt (~93%) was used extensively.

Complex hypophosphite salts have been used successfully as substitutes for both lead styphnate and tetracene. A 1939 patent gives the following composition:

Lead styphnate	33%
Calcium hypophosphite	7%
Lead nitrate	14%
Lead thiocyanate	10%
Barium nitrate	16%
Glass powder	20%

When wet with water a reaction occurs between the calcium hypophosphite and the lead nitrate, producing a shock-sensitive nonhygroscopic compound which incorporates both oxidizer and fuel.

In 1944 a patented rimfire priming mix included a triple salt, that is, basic lead styphnate–lead styphnate–lead hypophosphite, in the following mix:

Triple salt	50%
Lead nitrate	30%
Glass powder	20%

In 1955, patents were issued for a nontoxic, lead-free, rimfire priming mixture which used the double salt, ferric styphnate–ferric hypophosphite, and for a glassless rimfire priming mixture which used a triple salt, potassium styphnate–lead styphnate–lead hypophosphite, in the following unusual mixture:

Triple salt	10%
Lead styphnate	36%
Barium nitrate	50%
Tetracene	4%

About 1949 Frankford Arsenal manufactured an unusual priming mixture known as the P-4 primer (coded FA675):

| Stabilized red phosphorus | 18% |
| Barium nitrate | 82% |

Priming Compositions

Although this was a simple, relatively safe mixture, and was a satisfactory primer, it was discontinued after a very short period because of two major disadvantages. It was shown that copper, bismuth, silver, iron, and nickel increased the oxidation rate of red phosphorus to acidic compounds. Primer cups had to be zinc plated to prevent contact with copper. The red phosphorus had to be of high purity, and it was necessary to remove the major impurities (iron and copper) from commercial red phosphorus before use, and to coat the purified material with up to 7.5% aluminum hydroxide which inhibited oxidation.

Although the P-4 primer was only in use for approximately 1 year, it was further improved in 1961 by coating the stabilized red phosphorus with PETN, RDX (cyclotrimethylenetrinitramine), or TNT giving the following primer mix:

Stabilized red phosphorus	25%
PETN, RDX, or TNT	5%
Barium nitrate	70%

However, red phosphorus primers never achieved widespread use, presumably due to manufacturing difficulties.

In the early 1960s important advances were made in the development of safer, easier to make, cheaper, and better substitutes for lead styphnate, which had been the main explosive ingredient in successful NCNM priming mixtures up to this time.

In 1962 Kenney applied for patents on many complex, basic lead picrate-clathrate inclusion compounds which did not have the static electricity hazard of lead styphnate. Of 44 compounds listed in his patent, monobasic lead picrate–lead nitrate–lead acetate was preferred for primers, although monobasic lead picrate–lead nitrate–lead hypophosphite; dibasic lead picrate–lead nitrate–lead acetate; and monobasic lead picrate–lead nitrate–lead acetate–lead hypophosphite were also suitable. Glass was thought to damage the bore of the firearm and was considered by some to be undesirable. A glassless rimfire mixture was:

Any of the previous complex salts	46%
Barium nitrate	50%
Tetracene	4%

In 1962, Staba applied for patents on a double salt, lead nitroaminotetrazole–lead styphnate, which became known as stabanate, and had much better thermal stability than lead styphnate. A primer mix claimed to be superior to the lead styphnate–based equivalent was:

Stabanate	20%
Barium nitrate	50%
Antimony sulfide	15%
Tetracene	5%
Aluminum	10%

In 1966, Staba applied for a patent on certain forms of carbon that exhibit conchoidal fracture (very sharp, jagged concave edges) when shattered. A rimfire primer mix was:

Lead styphnate	20.00%
Stabanate	25.00%
Barium nitrate	36.25%
Tetracene	3.00%
Karaya gum	0.75%
Ground anthracite coal	15.00%

Another of Staba's primer mixes:

Stabanate	48.5%
Tetracene	3.0%
PETN	14.0%
Aluminum	10.0%
Nitroaminoguanidine	23.0%
Karaya gum	1.0%
Gum arabic	0.5%

An interesting stage in the development of primer mixes was the use of pyrophoric metal alloys, first patented in 1936 and improved in 1964. These rare earth alloys, as used in cigarette lighter flints, give a shower of sparks when lightly scraped. A typical pyrophoric alloy is "misch metal," which has the following approximate composition: cerium 50%, lanthanum 40%, other rare earth elements 3%, and iron 7%.

There are many patents listed in which the pyrophoric alloy replaces the function of both lead styphnate and tetracene. One of the most sensitive mixtures was:

Misch metal/magnesium (80/20 alloy)	50%
Barium nitrate	20%
Lead dioxide	10%
Zirconium powder	20%

Priming Compositions

Pyrophoric alloy primer mixtures never achieved widespread use, presumably because of their lack of sensitivity to percussion.[50]

There are hundreds of patents issued for priming compositions, a fact that illustrates the considerable experimentation in this area. Examples of some of these are:

Mercury fulminate	20% to 50%
Barium nitrate	19% to 45%
Lead chromate	2% to 20%
Lead sulfocyanide	3% to 25%
Zirconium powder	2% to 30%
Glass powder	30%
Basic lead trinitroresorcinol	27%
Lead dinitrophenylazide	13%
Potassium nitrate	30%
Antimony trisulfide	7%
Ground glass	23%
Mercury fulminate	33%
Thallium nitrate	40%
Cobalt nitrate	10%
Antimony trisulfide	17%
Potassium chlorate	85%
Asbestos fiber	1.5%
Nitrotoluol	4.5%
Petroleum gel	8.5%
Castor oil	0.5%
Potassium chlorate	48% to 53½%
Potassium ferrocyanide	33⅓% to 36%
Glass powder	13⅓% to 16%
Tetracene	1% to 4%
Diazonitrophenol	12% to 18%
Barium nitrate	25% to 40%
Antimony trisulfide	8% to 18%
Lead peroxide	15% to 25%
Calcium silicide	8% to 20%
Tetrazene	4% to 7%

Diazonitrophenol	15% to 20%
Basic lead azide	6% to 12%
Barium nitrate	20% to 30%
Lead peroxide	12% to 20%
Ground glass	20% to 28%
Lead azide	20 to 25 oz.
Powdered glass	20 to 25 oz.
Flake aluminum	6 to 8 oz.
Barium nitrate	35 to 38.5 oz.
Trinitrotoluol	0 to 25 oz.
Canada balsam or cellulose acetate	0 to 2.5 oz.
m-Toluenesulfomethylamide	0 to 1 oz.
Mercury fulminate	65.0 g
Barium nitrate	22.0 g
Antimony sulfide	11.0 g
Hexogene	15.5 g
Barium carbonate	1.5 g
Gum arabic	30 g
Phosphorus sulfide	15 g
Magnesium carbonate	12 g
Calcium carbonate	5 g
Potassium chlorate	60 g
Mercury fulminate	37.5%
Potassium chlorate	37.5%
Antimony sulfide	25.0%
Mercury fulminate	25.9%
Potassium chlorate	48.2%
Antimony sulfide	3.7%
Ground glass	22.2%
Mercury fulminate	19.1%
Potassium chlorate	33.3%
Antimony sulfide	42.8%
Sulfur	2.4%
Mealed powder	2.4%

Lead trinitroresorcinol	40%
Tetracene	2%
Barium nitrate	40%
Lead oxide	3%
Calcium silicate	11%
Powdered glass	4%

Despite the search for alternatives to lead styphnate and the considerable experimentation with primer compositions, in the United Kingdom and the United States, the vast majority of modern ammunition contains Sinoxyd type primers with lead styphnate and barium nitrate together typically making up 60% to 80% of the total weight. They also contain some of the following:

Antimony sulfide
Tetracene
Calcium silicide
Lead dioxide
Aluminum powder
Ground glass
Lead hypophosphite
Lead peroxide
Zirconium
Nitrocellulose
Pentaerythritol tetranitrate
Gum type binder

Composition control is very stringent and ingredients are of analytical reagent quality.

Mercury fulminate/potassium chlorate–based primer compositions are currently manufactured by some Eastern Bloc countries, although they also manufacture compositions based on lead styphnate.

Examples of some modern U.S. priming mixtures are[51]:

Normal lead styphnate	36%
Barium nitrate	29%
Antimony sulfide	9%
Lead dioxide	9%
Tetracene	3%
Zirconium	9%
Pentaerythritol tetranitrate	5%

Basic lead styphnate	39%
Barium nitrate	40%
Antimony sulfide	11%
Tetracene	4%
Nitrocellulose	6%
Normal lead styphnate	37%
Barium nitrate	38%
Antimony sulfide	11%
Tetracene	3%
Pentaerythritol tetranitrate	5%
Nitrocellulose	6%
Normal lead styphnate	41%
Barium nitrate	39%
Antimony sulfide	9%
Calcium silicide	8%
Tetracene	3%
Normal lead styphnate	43%
Barium nitrate	36%
Calcium silicide	12%
Tetracene	3%
Lead peroxide	6%

Examples of some modern U.K. priming mixtures are[52]:

Lead styphnate	35%
Tetracene	3%
Lead peroxide	15%
Barium nitrate	47%
Lead styphnate	46%
Tetracene	4%
Barium nitrate	25%
Antimony sulfide	20%
Aluminum	5%
Lead styphnate	44.2%
Tetracene	3.3%
Barium nitrate	20.4%
Ground glass	25.0%
Lead hypophosphite	6.8%

Gum arabic	0.3%
Lead styphnate	38%
Tetracene	2%
Lead peroxide	5%
Barium nitrate	39%
Antimony sulfide	5%
Calcium silicide	11%

An interesting and extremely successful primer innovation was introduced by Eley and is known as Eleyprime. Instead of using lead styphnate, with its inherent safety and processing difficulties, Eley uses calculated amounts of lead monoxide and styphnic acid which are much safer to process. At the end of the processing stage a drop of water is added to each individual primer which initiates a chemical reaction between the lead monoxide and the styphnic acid to form lead styphnate. The final product when dry is no different from a conventional primer.

In conventional ammunition lead, antimony, and barium are emitted when the ammunition is discharged. These three elements are undesirable from a health viewpoint and pose a major problem for firearms instructors in indoor firing ranges, as they are exposed to an unhealthy environment each working day. To solve this problem Dynamit Nobel AG developed a nontoxic primer composition called Sintox. Lead styphnate is replaced by 2-diazo-4,6-dinitrophenol (diazole) and the barium nitrate and antimony sulfide are replaced by a mixture of zinc peroxide and titanium metal powder.

The Sintox primer mixture contains tetracene, diazole, zinc peroxide/titanium powder, and nitrocellulose ball powder.[53] The use of this primer coupled with a totally jacketed bullet (base also enclosed) entirely eliminates the health hazard problem.

CCI and Fiocchi produce lead free primers, Fiocchi substituted diazole for the lead compound, and CCI uses diazole, manganese[iv] oxide, and aluminum.[54]

The use of titanium as a replacement for calcium silicide in conventional Sinoxyd primers is being investigated by Dynamit Nobel. Since they were introduced, lead free primers have improved to the extent that their performance rivals that of conventional lead-containing primers which they will probably replace in the near future.

Primer mixtures can be divided today into six categories: (a) mercuric and corrosive, (b) mercuric and noncorrosive, (c) nonmercuric and corrosive, (d) nonmercuric and noncorrosive, that is, Sinoxyd type, (e) Sintox type, that is, lead free, and (f) miscellaneous (unusual priming compositions).

Two-component primer compositions (based on lead and barium compounds) are more common than three-component types (based on lead, barium, and antimony compounds) in rimfire primed ammunition. However, three-component rimfire primers are far from rare. Some manufacturers use both two- and three-component primers in their range of rimfire ammunition.

Primers are not used exclusively for firearm ammunition, but have other uses which include blank cartridges, flares, flare trip wires, mortars, pyrotechnic cartridges, hand grenades, rocket-propelled grenades, ejector seat mechanisms, jettison devices, and other larger ammunition components.

References

45. *Kirk-Othmer Encyclopedia of Chemical Technology*, 2nd ed., 8 (New York: Wiley–Interscience), 654.
46. B. A. Bydal, "Percussion Primer Mixes," *Weapons Technology* (November/December 1971): 230.
47. G. R. Styers, "The History of Black Powder," *AFTE Journal* 19 (4) (October 1987): 443.
48. Dr. T. L. Davis, *Chemistry of Powder and Explosives* (New York: John Wiley & Sons), 42.
49. Major General J. S. Hatcher, *Hatcher's Notebook* (Stackpole Books), 353.
50. B. A. Bydal, "Percussion Primer Mixes," *Weapons Technology* (November/December 1971): 230.
51. J. E. Wessel, P. F. Jones, Q. Y. Kwan, R. S. Nesbitt, and E. J. Rattin, "Gunshot Residue Detection," The Aerospace Corporation, El Segundo, CA. Aerospace report no. ATR-75 (7915)-1 (September 1974), chap. 2, p. 13.
52. Private communication, 1975.
53. R. Hagel, and K. Redecker, "Sintox—A New, Non-Toxic Primer Composition by Dynamit Nobel AG," *Propellants, Explosives, Pyrotechnics* 11 (1986): 184.
54. W. Lichtenberg, "Methods for the Determination of Shooting Distance," *Forensic Science Review* 2, (1) (June 1990): 37.

Propellants   10

Small arms ammunition propellants may be defined as "explosive materials which are formulated, designed, manufactured, and initiated in such a manner as to permit the generation of large volumes of hot gases at highly controlled, predetermined rates."[55]

Ideally, a propellant would be a single, solid, nontoxic chemical compound that is stable, easy to store, easy to ignite, of compact mass, and so forth, which is cheap and simple to prepare from readily available materials and which on combustion produces no smoke or solid residue, that is, is completely converted into gas or gases. It must contain its own oxygen supply which is necessary for combustion in confined spaces, it must burn very rapidly as opposed to detonation, and it must have a satisfactory energy/weight relationship.

It is not surprising that no single chemical compound fulfills all these specifications. In practice propellants consist of a mixture of substances.

A propellant must fulfill the following general specifications:

It should be capable of being manufactured simply, rapidly, with relative safety, at reasonable cost and from ingredients that are readily obtainable in time of war (military propellants).

It must be easy and safe to load, nonhydroscopic, and free from combustion products that are difficult to remove or injurious to the firearm or cartridge case.

It must give consistent performance under varying conditions of storage and climate, and it must not deteriorate with age (this is especially applicable to propellants for military use which can be stored as a war reserve for a long period of time). It must also not ignite when in the chamber of a very hot firearm for a considerable period of time. (This also applies to priming compositions.)

The energy/weight/bulk relationship of a propellant and the rate of delivery of the energy must be matched to the system, that is, space available within the cartridge case and gun barrel, the bullet weight, pressure requirements, and the required bullet velocity. Consequently, a wide range of propellants is required to satisfy the varying ballistic requirements of a wide range of firearms and ammunition.

The burning rate is extremely important because if the propellant releases hot gases too quickly, it detonates, thereby destroying the gun and possibly causing injury to the firer. If it burns too slowly, it is inefficient, and the bullet will lack sufficient velocity. The burning rate can be controlled by the size and geometrical design of the individual granules. [An individual propellant particle is referred to as a grain (kernel, granule) and grains (kernels, granules) can be very small with simple geometries, or very large with complex geometries. Note that grains in this context should not be confused with the unit of weight; 7,000 grains = 1 lb = 453.59237 grams. In my opinion it is better to use the term granule to avoid confusion].

Apart from the inherent burning characteristics of a propellant the burning rate can also be varied by the use of surface coatings (moderants) on the granules of propellant.

Propellants are frequently referred to as gunpowder, powder charge, or simply as charge or powder. However, they are very rarely a true powder and are manufactured in a wide range of colors, shapes, and sizes. Figure 10.1 illustrates some shapes.

It is critically important that propellant granules not contain non-uniformities such as cracks, pores, and cavities, because this could cause internal granule burning, leading to detonation or excessive pressure.

The relationship between physical shape and burning rate is complex, dependent on the characteristics of the propellant surface, which affect the rate at which decomposition reactions occur, and also on the characteristics of the environment above the propellant, which affects the rate at which heat

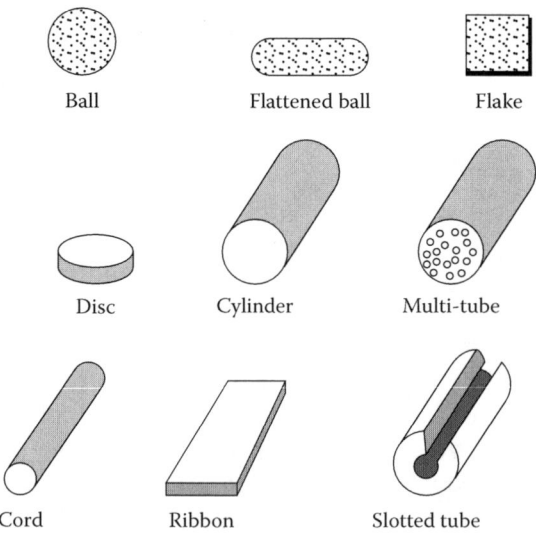

Figure 10.1 Propellant shapes.

is transferred to the propellant surface to cause chemical breakdown. Both surface and gas phase theories are intimately related.

The process of delivery of propellant gases at a predetermined rate involves the selection of a propellant composition with the required burning rate at the operating pressure of the firearm, and then designing the propellant granules so that the necessary burning surface is available to provide the required mass rate of gas evolution, that is, the necessary time/pressure relationship.

Since the introduction of smokeless powders in the period between 1870 and 1890, the use of black powder as a small arms ammunition propellant has substantially diminished. Black powder is still currently used as a propellant for some specialized purposes, for example, baton guns, punt guns, cable guns, signal flares, and by black powder firearms enthusiasts. It is also used in blank rounds of various types, and in many other ammunition components designed for larger caliber guns. Black powder suffers from several major disadvantages, namely, (a) a large amount of solid residue after combustion which attracts atmospheric moisture causing rusting of the firearm, (b) heavy fouling can also affect the efficient functioning of the firearms mechanism, (c) large amount of smoke formed after combustion can obscure the firer's view for subsequent shots, and (d) the smoke gives away the firing position.

Black powder is a mechanical mixture of charcoal, saltpeter (potassium nitrate), and sulfur in the typical proportion 15:75:10, respectively. The charcoal is the fuel, the saltpeter supplies the oxygen necessary for combustion in a confined space, and the sulfur is a binding agent that aids in holding the mixture together and to a much lesser extent also acts as a fuel. Black powder is black and granular in appearance and the burning rate is controlled by granulation size.

When black powder burns, the "initial portion" ignited undergoes a chemical reaction which results in the production of hot gases. The gases expand in all directions, warming the next portion to the "kindling" temperature. This then ignites, producing more hot gases and raising the temperature of the next portion, and so on. As the black powder is confined in the cartridge case the pressure rises and the heat cannot escape; consequently, it is communicated rapidly throughout the mass. In a confined space the combustion becomes extremely rapid; consequently, the pressure rise is also extremely rapid.

Black powder burns to produce a dense white smoke which contains extremely small particles held temporarily in suspension by the hot combustion gases.

Analysis of the combustion products of a particular brand of black powder gave the following results[56]: 42.98% of its weight as gases, 55.19% solids, and 1.11% water. Analysis of the solid products (percent by weight) and of the gaseous products (percent by volume) is as follows:

Solid Products		Gaseous Products	
Potassium carbonate	61.03	Carbon dioxide	49.29
Potassium sulfate	15.10	Carbon monoxide	12.47
Potassium sulfide	14.45	Nitrogen	32.91
Potassium thiocyanate	0.22	Hydrogen sulfide	2.65
Potassium nitrate	0.27	Methane	0.43
Ammonium carbonate	0.08	Hydrogen	2.19
Sulfur	8.74		
Carbon	0.08		

Black powder can vary from brand to brand. Variations in percentage compositions between manufacturers are small, but different charcoals, types of saltpeter (purity), different moisture content, and so forth can result in different ballistic performances from basically similar mixtures. Owing to a temporary shortage of potassium nitrate during World War I, sodium nitrate was used as a substitute. Ammonium nitrate has also been used as a substitute for potassium nitrate.

Brown powder (cocoa powder) represents the peak of development of black powder and was the most successful form of black powder exhibiting better burning characteristics. It was made in single perforated hexagonal or octagonal prisms. A partially burned brown charcoal made from rye straw, which had colloidal properties and flowed under pressure, cementing the granules together, was used. This made possible the manufacture of slow burning propellant containing little or no sulfur. A typical brown powder was brown charcoal 19%, saltpeter 78%, and sulfur 3%. A sulfur-free brown powder was brown charcoal 20% and saltpeter 80%.

A modern substitute for black powder is "Pyrodex." It is safer to transport, store, and use, and is cleaner burning than conventional black powder. Pyrodex incorporates both charcoal and sulfur but in much smaller proportions than in black powder, and potassium nitrate in addition to other ingredients. Pyrodex also contains potassium perchlorate, sodium benzoate, and dicyandiamide.[57]

Modern smokeless propellants for small arms ammunition almost exclusively contain plasticized cellulose nitrate (NC) as the major oxidizing ingredient (cellulose hexanitrate, commonly referred to as nitrocellulose). Various other chemicals are added for specific purposes:

> High energy oxidizing plasticizers such as nitroglycerine (NG–glyceryl trinitrate) to increase performance.
> Fuel type plasticizers such as phthalates, polyester adipate, or urethane to improve physical and processing characteristics.

Propellants

- Organic crystalline chemicals such as nitroguanidine to moderate the ballistic characteristics.
- Stabilizers such as diphenylamine, 2-nitrophenylamine, dinitrotoluene, N-methyl-p-nitroaniline, centralites, or acardites (e.g., N,N¹-diphenylurea), to increase chemical stability by combining with decomposition products.
- A range of inorganic additives such as chalk, graphite, potassium sulfate, potassium nitrate, barium nitrate, to improve ignitability, facilitate handling, and minimize muzzle flash. (Graphite acts as a lubricant to cover the granules and prevent them from sticking together and it also helps to dissipate static electricity).
- Powdered metals are sometimes added to change thermal characteristics such as conductivity.
- Some manufacturers also add colored taggants to aid in identifying their product.

Propellants that contain nitrocellulose as the only oxidizer are referred to as single base and propellants that contain both nitrocellulose and nitroglycerine (or some other explosive plasticizer) as double base. Triple-based propellants are produced when substantial quantities of an organic, energy-producing, crystalline compound such as nitroguanidine are incorporated in double-based propellants. Triple-based propellants are unlikely to be encountered in small arms ammunition.

Stabilizers are necessary because nitrocellulose decomposes with age. The decomposition reaction yields dinitrogen tetraoxide which acts as an autocatalyst and accelerates the decomposition.[58] Stabilizers act as dinitrogen tetraoxide scavengers; consequently shelf life is increased. Stabilizers are normally added in the region of 0.5 to 2.0%. To neutralize the decomposition products, which could cause corrosion of the firearm, calcium carbonate is added to some propellants. A common stabilizer is diphenylamine or its nitro derivatives (Figure 10.2).

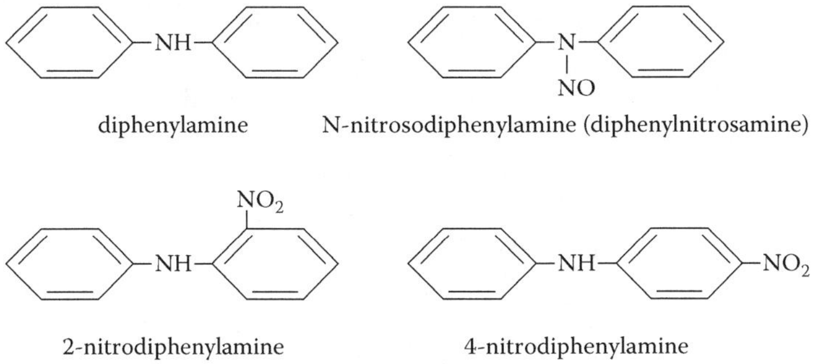

Figure 10.2 Stabilizers.

```
       C₂H₅      C₂H₅
        |         |
  ⌬—N—C—N—⌬
            ‖
            O
```

Figure 10.3 Ethyl centralite (smydiethyl diphenylurea).

Diphenylamine is the most common stabilizer especially in single-based powders. It has been suggested that diphenylamine is not a good stabilizer for double-based propellants as it may hydrolyze NG.[59] 2-Nitrodiphenylamine is used for double- and triple-based propellants.

Another common stabilizer is ethyl centralite (Figure 10.3) although methyl centralite is sometimes used.[60] Methyl centralite (Sym-dimethyl diphenylurea; Centralite II) is also used as a moderant to reduce the burning rate. Ethyl centralite is usually found in double-based propellants. Resorcinol (Figure 10.4) is also used as a stabilizer.

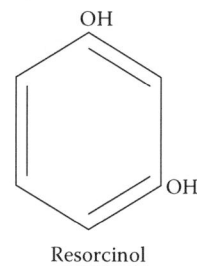

Resorcinol

Figure 10.4 Resorcinol.

Plasticizers add strength and flexibility to the propellant granules. Examples of some plasticizers used are shown in Figure 10.5 and Figure 10.6.[61,62]

Muzzle flash suppressors (flash reducers) include dinitrotoluene (Figure 10.7). Dinitrotoluene acts as a flash suppressor by reducing the heat of explosion. Nitroguanidine (picrite) is another flash suppressor which acts by producing nitrogen, thereby diluting the combustible muzzle gases. Potassium nitrate and potassium sulfate are also used as flash suppressors but both have the disadvantage of producing smoke.

Wear reduction additives include wax, talc, and titanium dioxide.

Binders (to hold the granule shape) include ethyl acetate, and rosin (also called colophony; the sap or sticky substance from pine or spruce trees).

Decoppering additives used to decrease the buildup of copper residues in the barrel rifling include tin metal and compounds such as tin dioxide; bismuth metal and compounds such as bismuth trioxide, bismuth subcarbonate, bismuth nitrate, bismuth antimonide. The bismuth compounds are

```
CH₂—O—COCH₃
 |
CH—O—COCH₃     Triacetin (Glyceryl triacetate)
 |
CH₂—O—COCH₃
```

Figure 10.5 Triacetin (glyceryl triacetate).

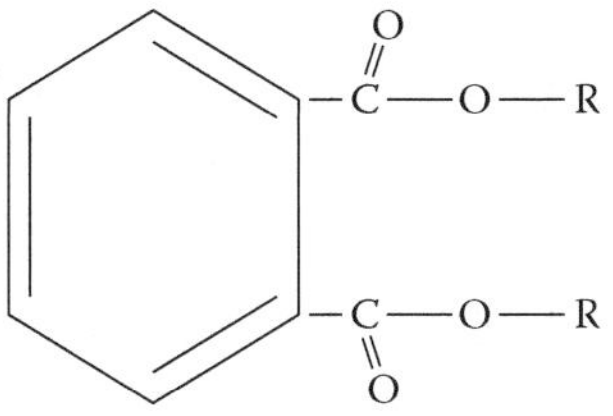

Propellants

R is CH$_3$ — dimethyl phthalate
R is C$_2$H$_5$ — diethyl phthalate
R is C$_4$H$_9$ — dibutyl phthalate

Figure 10.6 Dimethyl phthalate, diethyl phthalate, and dibutyl phthalate.

preferred as copper "dissolves" in molten bismuth, forming brittle and easily removable alloy. Lead foil and compounds were also used but due to toxicity they are being phased out.

Examples of single- and double-based propellant compositions are given in Table 10.1 and Table 10.2.

Smokeless powders leave relatively little solid residue on combustion and produce much less smoke than black powder. Combustion of smokeless powders produces primarily nitrogen, carbon monoxide, carbon dioxide, hydrogen, and water vapor. The quantity of smokeless powder varies depending on the caliber, bullet weight/type, required pressure/velocity, space available within the cartridge case/chamber, and so forth. Ammunition for use in rifles contains propellant varying in weight from ~0.45 g (6.9 grains) for a .22" caliber to ~6.45 g (99.5 grains) for a .378" caliber. For pistols/revolvers the range can vary from ~0.06 g (0.9 grains) for a .25" caliber to ~1.72 g (26.5 grains) for a .44" Magnum caliber. For shotguns the range can vary from ~1.10 g (17.0 grains) for a 20-bore caliber to ~2.0 g (30.9 grains) for a 12-bore caliber.

Generally, about 700 to 1,100 cm^3 of gas per gram is produced and flame temperature can range from, for example, 2,000 K for a cool propellant to 4,000 K for very hot propellants. Typical gas composition from double-based

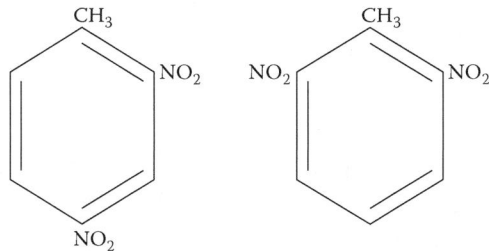

Figure 10.7 2,4-Dinitrotoluene and 2,6-dinitrotoluene.

Table 10.1 Single-Based Propellants (% composition)

Nitrocellulose (NC)	89.0	99.0	97.7	90.0	79.0	85.0	87.0	96.25	94.25	94.0	98.0	99.4	92.4
Barium nitrate	6.0						6.0						
Potassium nitrate	3.0						2.0				.25		
Starch	0.75												
Paraffin oil							4.0						
Diphenylamine	1.0	1.0	0.8	1.0	1.0	1.0	1.0	1.0	1.0	1.0	1.0	0.6	0.6
Dinitrotoluene				8.0		10.0							6.5
Methyl centralite								2.0	2.0	1.75			
Dibutyl phthalate				1.0		4.0			2.0	1.75			
Glyceryl triacetate					5.0								
Tin			0.75							0.8			
Graphite					0.2			With NC	With NC		0.5		0.5
Potassium sulfate			0.75					0.75	0.75	0.75			
Dye (aurine)	0.25												
Trinitrotoluene					15.0								

Table 10.2 Double-Based Propellants (% composition)

Nitrocellulose	77.45	52.15	56.50	59.65	85.45	59.40	58.00	76.5	89.4	79.25	51.5
Nitroglycerine	19.50	43.00	28.00	36.00	9.00	36.00	40.00	21.5	8.0	15.00	43.00
Diethylphthalate		3.00								3.50	3.25
Dibutylphthalate				0.40	0.40	0.40					
Diphenylphthalate					1.10						
Dinitrotoluene			11.00	0.35	0.65						
Potassium sulfate		1.25	1.50*						0.8		1.3
Potassium nitrate	0.75										
Ethyl centralite	0.60	0.60	4.50					2.0	1.0		1.00
Graphite	0.30			1.05	0.25	1.00	0.60			0.60	0.20*
Barium nitrate	1.40										
Candelilla wax			0.08*								
Methyl cellulose			0.50*								
Sodium sulfate				0.10	0.10	0.10	0.10			0.10	
Calcium carbonate				0.40	0.40	0.40	0.10			0.10	
Diphenylamine				1.00	1.00	1.00	0.50			0.50	
Water				0.50	0.90	0.55	0.40			0.60	
Methyl centralite									5.0*		
Tin									0.8		

* Added to basic composition.

propellants is carbon dioxide 28%, carbon monoxide 23%, hydrogen 8%, nitrogen 15%, and water 26%.

Other ingredients that may be found in smokeless powders include camphor, carbazole, cresol, diethyleneglycoldinitrate (DEGDN), dimethylsebacate, dinitrocresol, di-normal-propyl adipate, 2.4-dinitrodiphenylamine, PETN, TNT, RDX, acaroid resin, gum arabic, synthetic resins, aluminum, ammonium chloride/oxalate/perchlorate, pentaerythritol dioleate, oxamide, lead carbonate/salicylate/stearate, magnesium oxide, sodium aluminum fluoride, sodium carbonate/bicarbonate, petrolatum, dioctylphthalate, stannic oxide, potassium cyrolate, triphenyl bismuth.

The percentage of NG in double-based propellants can range from as low as 5% to as high as 44%.

Apart from firearms ammunition other propellant-activated devices have numerous uses, for example, to drive turbines, to move pistons, to eject pilots from jet planes, to shear bolts and wires, to operate vanes in rockets, to act as sources of heat in special operations, to operate pumps in missiles, to clear blocked drill bits underground, to start aircraft engines, to jettison stores from aircraft, and generally for systems that require well-controlled sources of high force applied over relatively short periods of time. Propellants are also used in some blank cartridges.

References

55. *Kirk-Othmer Encyclopedia of Chemical Technology,* 2nd ed., vol. 8 (New York: Wiley-Interscience), 659.
56. Dr. T. L. Davis, *Chemistry of Powder and Explosives* (New York: John Wiley & Sons), 43.
57. E. C. Bender, "The Analysis of Dicyandiamide and Sodium Benzoate in Pyrodex by HPLC," *Crime Laboratory Digest* 16, no. 3 (October 1989): 76.
58. T. Urbanski, *Chemistry and Technology of Explosives,* vol. 3 (Oxford: Pergamon Press, 1967), 554.
59. Urbanski, *Chemistry and Technology of Explosives,* 561.
60. Urbanski, *Chemistry and Technology of Explosives,* 645.
61. S. Fordham, *High Explosives and Propellants,* 2nd ed. (Oxford: Pergamon Press, 1980), 171.
62. J. M. Trowell, and M. C. Philpot, "Gas Chromatographic Determination of Plasticizers and Stabilizers in Composite Modified Double Base Propellant," *Analytical Chemistry* 41 (1969): 166.

Projectiles   11

The Firearms Act (Northern Ireland) defines a firearm as "a lethal barreled weapon of any description, from which any shot, bullet or other missile can be discharged." This very loose definition leaves scope to cover almost every conceivable type of device that incorporates a tube through which any missile is projected. Could a blowpipe used to discharge poisoned darts be described as a firearm? It is "gas" operated, has a barrel, and the projectile is designed to be lethal. Although the blowpipe may not be considered a lethal barrel, it is, as with a firearm, the means of directing and discharging the projectile. It is the projectile that kills. Consequently, great attention has been focused on projectile design, and there are many different types of projectiles available on the military and civilian markets.

For the purpose of this text, only conventional projectiles are considered in detail. Conventional projectiles for firearms are bullets, pellets, and slugs, each of which may differ from others of the same kind in size, shape, weight, composition, and physical properties.

There is a wide range of firearms, and the choice of ammunition available presents a very large number of gun/ammunition combinations. The reason for such a variety of projectiles encompasses internal and external ballistics, nature of target, and wound ballistics, all of which are beyond the scope of this text.

Bullets

Every bullet type is designed for a specific purpose and the range of bullet designs available for a single firearm can be substantial. Figure 11.1 illustrates some different physical designs of round nose bullets.[63] This only illustrates a variation of types within one design of bullet that is available in a range of calibers. Variation of types occurs within other designs of bullet, for example, truncated cone, cone- or spire-point, spitzer, flat nose, semi-wad-cutter, wad-cutter, round ball, and so forth, all of which are available in a range of calibers. Even the design of the base of the bullet can vary.[64] This is illustrated in Figure 11.2.

There is also a wide variety of bullet core/bullet jacket designs without even considering compositional differences. Bullets are unjacketed, or

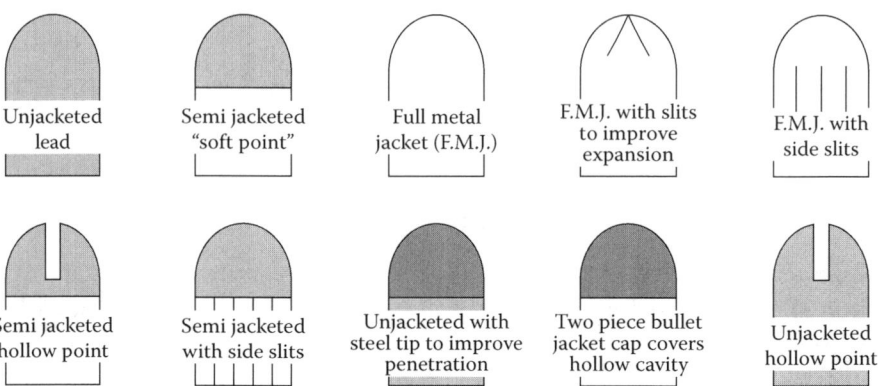

Figure 11.1 Designs of round-nosed bullets.

jacketed (envelope), or partially jacketed. Unjacketed bullets are usually confined to revolvers or low-power pistols and rifles. Such bullets may have their surface coated with a very thin layer of copper or brass colored material which is used as a lubricant, and for cosmetic reasons. This is referred to as a "coat" or "wash" and is not a bullet jacket in the conventional sense of the word. Unjacketed bullets are frequently lubricated with some form of wax or grease to prevent or reduce lead fouling in the barrel of the firearm.

Higher velocity bullets have to be either full or partly jacketed because an unjacketed lead bullet fired at high velocity can suffer deformation and have a detrimental amount of lead stripped from its surface by the rifling grooves. Such lead deposited inside the barrel has a pronounced effect on accuracy of subsequent shots. Unjacketed lead bullets are also prone to damage by the feeding mechanisms of self-loading firearms.

In the majority of bullets the lead base is exposed to the hot propellant gases. This applies to unjacketed and jacketed bullets (excluding total metal jacketed bullets). Some bullets incorporate a gas check in the base to prevent erosion by the hot gases. Such erosion can upset the symmetry of the bullet and consequently the accuracy. The base of the bullet may be filled or covered with a substance, for example, Alox base lubricant, that is unaffected by the temperature and pressure generated during discharge. Another method is to enclose the base with a shallow copper cup. Some bullets have the base enclosed by the jacket. Electroplated jackets usually cover the entire bullet and some soft point bullets with a nose of exposed lead have a partial jacket which is usually closed at the base. Bullets that are totally

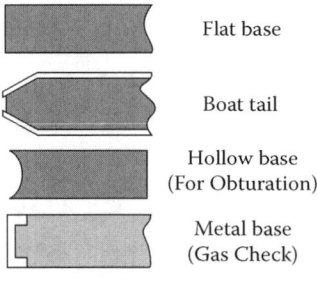

Figure 11.2 Bullet base designs.

enclosed by the jacket, including their base, are referred to as total metal jacketed bullets (TMJ).

Conventional bullets are referred to as ball loads, the word "ball" originating from the use of round balls as projectiles in the early days of firearm development. Modern bullets are nonspherical projectiles for use in rifled barrels. Conventional bullets are designed for either penetrating power or stopping power (transfer of all kinetic energy on impact thereby rapidly stopping the human or animal target) or a combination of both. This is achieved by physical design and the selection of materials with suitable physical properties.

The bullet jacket material is almost always harder than the bullet core material, with the one exception of armor-piercing bullet jackets. Bullet jacketing is done either by electroplating or, much more commonly, the jacket is manufactured separately from the bullet, and the bullet then forced into the jacket in a press. Another method is to pour molten lead into the jacket. The edges of the jacket are usually partly rolled over the base of the bullet or attached by some other physical means.

Whenever a jacketed bullet strikes a target it is possible for the core and jacket to separate, with a consequent reduction in penetration. To prevent such an occurrence a variety of crimps, folds, jacket geometries, and melted core techniques are employed. Another method of interest to hand loaders is the use of a product called Core-Bond which is a flux that removes surface oxides allowing molten lead to bond directly to the jacket. This allows a degree of alloying between the two metals, which is claimed to provide bonding superior to that achieved by physical methods. Soldering of the jacket to the core has also been employed.[65]

Bullet jacket materials include gilding metal; cupronickel; cupronickel-coated steel; nickel; zinc-, chromium-, or copper-coated steel; lacquered steel; brass; nickel- or chromium-plated brass; copper; bronze; aluminum/aluminum alloy; Nylon (Nyclad), Teflon- and cadmium-coated steel (rare). Black Talon bullets have a black molybdenum disulfide coating over the metal bullet jacket which acts as a dry lubricant. Steel jackets are frequently coated both inside and outside as an anticorrosion measure. Gilding metal is by far the most common bullet jacket material. Tin is claimed to have lubricating properties and is sometimes incorporated in bullet jacket material. The alloy is known as Lubaloy or Nobaloy and contains 90% copper, 8% zinc, and 2% tin.

The thickness and hardness of the jacket can vary between the base and the nose of the bullet, with the nose portion thinner for better expansion on impact or thicker for greater penetration of the target. The way the jacket is physically attached to the core can vary. This depends on the desired effect of the bullet on the target, either the controlled expansion of the bullet, greater penetration of the bullet, or the prevention of core and jacket separation. Figure 11.3 illustrates some different physical designs.[66]

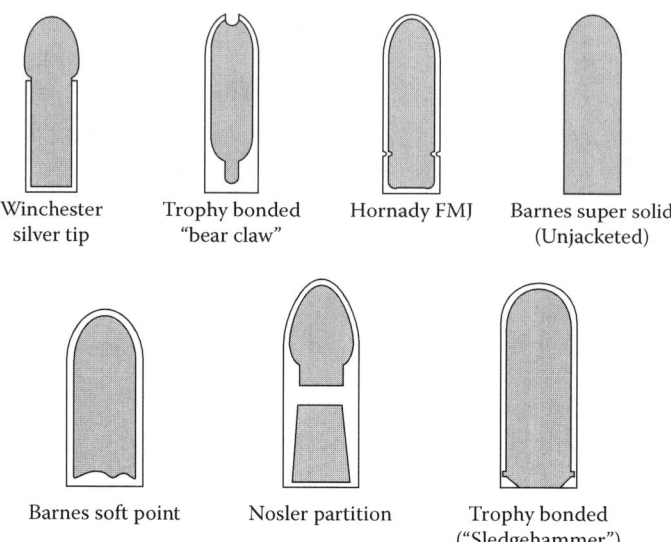

Figure 11.3 Bullet core/jacket designs.

The core of the bullet can be made from a variety of materials; lead is by far the most common because of its high density and the fact that it is cheap, readily obtained, and easy to fabricate. But copper, brass, bronze, aluminum, steel (sometimes hardened by heat treatment), depleted uranium, zinc, iron, tungsten, rubber, and various plastics may also be encountered. (When most of the fissile radioactive isotopes of uranium are removed from natural uranium, the residue is called depleted uranium. Depleted uranium is 67% denser than lead, and it is an ideal bullet material and is very effective in an armor-piercing role, both in small arms and larger munitions components. Because of its residual radioactivity its use is controversial.) Bullets with a lead core and a copper alloy jacket are by far the most common.

Sometimes a combination of bullet core materials is used to produce a hardness difference between the base and the nose (dual core bullets), for example, jacketed bullets with a lead nose and a steel base, a steel nose and a lead base, or a soft lead nose and a hardened lead base.

Bullet lead can be either soft lead or lead hardened by antimony, by tin, or by both. Mercury was also used to harden lead in the early days of bullet development. The quantity of alloying materials varies considerably, for example, antimony <0.5% to as high as 12% but typically 2% to 5%, tin <0.5% to 10% but typically 3% to 5%. A larger amount of tin is required to give the same degree of hardness as that of antimony; consequently, for cost reasons, antimony is more frequently used.

Some jacketed bullets incorporate a small cavity in the nose which is filled with a material different from the bullet core. In some bullets the cavity is unfilled. Bullet tips are usually made from lighter-weight material than that used for the core, for example, plastic, aluminum, fiber, sodium carbonate, polycarbonate, nylon, paper, mild steel. Some soft point and unjacketed bullets employ a metal cap over the nose of the bullet either for increased penetration, to protect the nose from damage, or to improve feeding in self-loading firearms. Copper, steel, aluminum, and plastic caps are encountered.

Non-ball bullet types are often referred to as "filled bullets" and are designed for special purposes, for example, tracer, incendiary, and armor-piercing roles.

Armor-Piercing Bullets

Armor-piercing (AP) ammunition has a projectile or projectile core constructed entirely from a combination of tungsten alloys, steel, iron, brass, bronze, beryllium copper, or depleted uranium. The most effective AP bullets are usually confined to rifle bullets, as velocity and range are important factors in AP requirements. Some revolver and pistol ammunition is described as metal piercing but, although it would be effective against vehicle bodywork and some body armor, it would be ineffective against "heavy" armor plate. AP bullets are, with very few exceptions, jacketed.

AP rifle bullets usually have a bullet tip filler (usually lead) which is designed to "cushion" the effect of the impact on the AP core, which is very hard and brittle and can break on impact without a "cushioning" effect. The AP core is also frequently surrounded with a thin sheath of lead between the core and the bullet jacket. The AP core is usually hardened steel such as tungsten/carbon, tungsten/chromium, manganese/molybdenum, chromium/vanadium, or chromium/molybdenum.

Two specifications for AP core material are:

1. Steel with 2% to 3% tungsten and 1.25% carbon
2. Steel with 3% to 4% tungsten, 1.1% carbon, and a trace of manganese[67]

Tungsten carbide has also been used as an AP core and gives much superior penetration. It is produced by alloying tungsten with carbon, nickel, cobalt, or other elements. Tungsten carbide is about twice as dense as steel and about 1.4 times denser than lead. It is nonmagnetic and extremely hard. Analysis of such a core yielded the following results:

Tungsten	93.90%
Carbon	1.65%
Titanium	1.55%
Nickel	1.55%
Iron	0.43%
	+ trace elements

A tungsten carbide AP bullet which also incorporates a lachrymal agent and a tungsten carbide AP core/tracer/tear gas bullet have been manufactured. A similar AP/tracer/tear gas bullet was also manufactured with hardened steel, rather than a tungsten carbide core.

There is a Chinese tungsten carbide AP bullet with a discarding sabot which is very effective due to its very high velocity. The United States also produced an AP bullet with a discarding sabot using depleted uranium as the bullet core material.[68]

Revolver and pistol metal-piercing bullets are available in a range of calibers and designs:

> KTW bullet: The original design had a hard steel or tungsten steel core with a copper gas check and the current version is a solid brass or bronze bullet without a gas check. Both have a gliding metal half jacket and the exposed portion of the bullet is coated with green-colored Teflon.
> National bullet: A sharply pointed solid steel bullet the base of which is contained within a brass cup.
> ABC (American Ballistic Company) bullet: A solid steel-pointed bullet without a jacket.
> Arcane bullet: A cone-shaped flat base bullet made from solid copper alloy (the .380" caliber is round nosed).
> THV bullet (Tres Haute Vitesse): An unusually shaped bullet made of solid brass with three times more penetrating power than a conventional bullet due to its hardness, geometric design, and high velocity.

Tracer Bullets

When fired, a tracer bullet leaves a visible trace behind it so that the trajectory can be seen and the aim corrected if necessary (trajectory tracer). A tracer bullet has a cavity at the base which is filled with a mixture of substances that are ignited by the propellant. The bullets are available in a range of rifle, pistol, and revolver calibers, although they are more common in rifle calibers.

The tracer composition is frequently housed in a metal canister placed inside the base cavity. Copper, brass, gilding metal, gilding metal-coated steel, and copper-coated steel canisters are known. Some compositions are placed into the hollow base cavity without the use of a canister. A typical tracer bullet consists of a gilding metal bullet jacket and a lead core with a base cavity containing the tracer composition. Paper discs and lead, steel, or brass washers sealed with varnish are sometimes used to "seal" the base of the bullet.

Four examples of tracer compositions are[69]:

> Barium peroxide 86% (oxidant and coloring agent)
> Magnesium powder 12% (combustible material)

Acaroid resin	2% (binder)
Strontium nitrate	51.7% (oxidant and coloring agent)
Magnesium	33.3% (combustible material)
Polyvinylchloride	5.4%
Phenol formaldehyde (with yellow dye)	9.6%
Magnesium powder	38.0% (combustible material)
Beeswax	4.8%
Strontium nitrate	42.8% (oxidant and coloring agent)
Shellac	4.8%
Chlorinated rubber	4.8%
Magnesium carbonate	4.8%
Magnesium	13% (combustible material)
Aluminum	3%
Strontium nitrate	73% (oxidant and coloring agent)
Iron	6%
Resinous material	5% (binder)

Dark or dim ignition tracer bullets are designed to "ignite" some distance from the muzzle to avoid dazzling the firer and giving the firing position away. They are also known as delay tracer bullets. These bullets have a duller, slower burning igniter composition in the base above which is the bright burning tracer composition. One tracer igniter mixture consists of potassium permanganate and iron filings.[70] Titanium metal is also used in some tracer compositions.

There are night tracer, day tracer, and dim ignition tracer bullets. Yellow, green, orange, red, and white tracer colors are known and they can be manufactured in different degrees of luminosity. Tracer igniter compositions also have varying degrees of luminosity, that is, bright for daytime use and dim or dark for nighttime use.

An explosive/tracer rifle bullet is known and consists of a copper alloy bullet jacket and a steel core. There is an explosive charge at the top of the cavity which consists of 40% PETN, 45% lead azide, and 15% tetracene. Below the explosive charge is black powder or smokeless powder contained in a small metal cup followed by the tracer composition at the base.

Tracer bullets can also be used to indicate when the ammunition in a magazine or belt is nearly exhausted by putting tracer bullets in a known sequence with ball or other ammunition.

Incendiary Bullets

Incendiary bullets are used against flammable targets and have an incendiary composition in the nose portion which is ignited on impact. Incendiary bullets are usually encountered in rifle calibers.

The best-known incendiary agent is magnesium, which melts about 650°C and once melted is very easily ignited. Incendiary compositions capable of being easily ignited and which can evolve enough heat to melt the magnesium are used. Some examples of incendiary compositions are[71]:

Aluminum powder	20%
"Thermite" Iron oxide (Hammerscale)	40%
Barium nitrate	35%
Boric acid	5%

This mixture is unlikely to be encountered in small arms ammunition. The following two mixtures are more likely to be encountered:

Magnesium/aluminum alloy	50%
Barium nitrate	50%
Barium nitrate	32.0%
Barium peroxide	53.3%
Magnesium	9.8%
Phenol formaldehyde	4.9%

Phosphorus is another well-known incendiary agent, and phosphorus or phosphorus-based compositions have been used in both incendiary and tracer rounds. Because of manufacturing difficulties in working with phosphorus, such compositions have largely been replaced with magnesium type mixtures. However, they are encountered in older ammunition and are still manufactured in some non-European countries. Although presenting manufacturing difficulties, phosphorus or phosphorus-based compositions were effective both in incendiary and tracer roles. One such incendiary composition was a mixture of phosphorus and aluminum.

Other incendiary compositions used in older ammunition were potassium chlorate based with a mixture of potassium chlorate and mercury sulfocyanide as the "priming" composition. Another older incendiary composition consisted of potassium nitrate, magnesium, aluminum, and lead oxide.[72,73] Multipurpose filled bullets are also manufactured, for example, armor piercing/incendiary, armor piercing/tracer, and spotter tracer bullets, which leave a visible trace and produce a puff of smoke on impact. One such smoke charge is lead dioxide 85% and powdered aluminum 15%. There

Projectiles

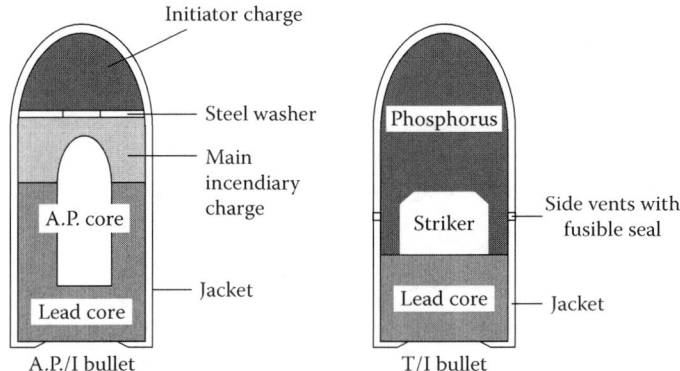

Figure 11.4 Multipurpose filled bullets.

are also tracer/incendiary and armor piercing/incendiary/tracer bullets. An explosive charge or a lachrymal agent may also be incorporated in multipurpose filled bullets. Two World War II examples of multipurpose filled bullets are shown in Figure 11.4.

In the armor-piercing/incendiary bullet the initiator charge was a mixture of potassium chlorate, magnesium, and antimony. The main incendiary composition was a mixture of magnesium, aluminum, barium chlorate, sulfur, and nitrocellulose flake powder. The tracer/incendiary bullet had two side vents which were sealed with a fusible metal consisting of tin, bismuth, and antimony. On passage down the barrel, the frictional heat melted the fusible seal and ignition of the phosphorus occurred on contact with the air, enabling the phosphorus to act as a tracer as well as an incendiary.

Shotgun Pellets and Slugs

Shotgun pellets (shot) are round metal balls available in various sizes, degrees of hardness, and materials depending on their intended use. An unusual cubic shot load is manufactured in 12, 16, and 20 gauge and is claimed to give a faster opening pattern and reduced ricochet hazard than do round pellets.

In shotgun cartridges the pellets are contained within the cartridge case. The pellets in a cartridge are normally the same size and composition, but cartridges are available that are loaded with a range of different pellet sizes (Duplex).

Shotgun pellets can be soft shot, that is, lead containing less than 0.5% alloying metal, or hard shot (chilled shot), that is, lead containing between 0.5% and 2.0% alloying metal, or extra hard shot, that is, lead containing 3% or more alloying metal, or steel shot (iron shot), that is, soft steel pellets. (Steel shot was introduced for environmental reasons, as lead is toxic. However, steel shot has proved to give performance problems, leading to a search for a suitable nontoxic alternative. (Bismuth shot is being evaluated and looks promising in terms of both performance and toxicity level.)

The pellets are sometimes graphite coated, copper plated, or nickel plated. Plated shot is normally confined to hard or extra hard shot. The alloying metal is usually antimony, and lead pellets typically contain between 0% and 6% antimony, although up to 12% antimony has been encountered. Tungsten–polymer shot is also manufactured and, as the name implies, contains tungsten metal embedded in some sort of thermoplastic.

Sometimes a plastic buffer material (usually white granulated polyethylene) is mixed with the shot to help prevent distortion during discharge. The buffer material may be found in the wound channel in close range shootings. Another innovation to prevent distortion is the use of a shot cup that encloses the shot as it travels through the barrel, with the shot cup falling away when it exits the muzzle.

Up to four or five wads can be present in a shotgun cartridge, although two or three are more common. The wads are used to provide a gas tight seal as the projectiles pass into the barrel, to separate the propellant from the pellets, and to close the mouth of the cartridge case. Wads are usually made from paper or cardboard, plastic (range of colors), felt (which may or may not be waxed or greased), and occasionally cloth type material. Wads may be faced with waxed paper or by black or colored glazed paper, and the overshot wad may be covered with a paper label containing printed information about the cartridge.

On discharge, the wads are propelled from the muzzle, but due to their shape and weight they only travel a short distance unless carried by the wind. However, in close range shootings (usually less than 6 feet) a wad or wads can be encountered in the wound channel.

In shotgun shootings, the wads can provide useful information. They can be chemically examined to provide some information about the primer and propellant types, and physical examination can sometimes reveal the caliber and give an indication of shot size. Wads from different manufactures can vary in composition, color, thickness, and design, and physical examination may reveal the identity of the cartridge. (Wads are also found in some rifle, revolver, and pistol ammunition. There is never more than one wad and such ammunition is rare. Sometimes the wad is made of nitrocellulose but more commonly of glazed or waxed cardboard. Such wads are placed between the propellant and the bullet.)

Sometimes a single rifled slug, a single lead ball, or a plated steel ball is loaded in shotgun cartridges. These are designed for heavy game at short range or for special purposes, for example, by the police to penetrate an engine block, thereby disabling a fleeing vehicle. Rifled slugs transform the shotgun into a makeshift short range rifle capable of considerable penetration and great stopping power at close range. Some slugs are made with hollow points to increase expansion on hitting the target. A sabot slug is manufactured which can be fired from a smooth bore but is most effective when fired from a rifled shotgun barrel. Special purpose ammunition is available for use in shotguns, for example, large slugs for use against door locks and hinges for forced entry into premises and vehicles; saboted bullets and slugs, tear gas, pepper balls, smoke screen; flechette or rubber pellet loads and beanbags for riot control. Special shotgun loads are also used by the military for bomb disposal work. Shotgun cartridges, mainly because of their larger size, are particularly suitable for loading with special devices and chemical agents.

Several examples of special loads are as follows. A single cylindrical frangible metal/ceramic slug which appears to be a mixture of iron and dental plaster is used in the "Shok-Lock" round which is designed for gaining entry to dwellings by destroying door locks. Sometimes a wad is attached to the base of a slug to provide a gas tight seal as it travels through the barrel of the shotgun.

Another unusual 12-bore shotgun round is the French "Silver Plus." It is an arrow-shaped bullet fired from a shotgun and is claimed to give increased velocity, four times the accuracy, and three times the penetration of a conventional lead slug. The bullet is surrounded by a two-piece ring sleeve, the two halves of which fall away whenever the projective leaves the muzzle.[74]

Hand Loading

Hand loading (making ammunition from new component parts, namely, cartridge case, primer, propellant, and bullet) and reloading (making ammunition by reusing spent cartridge cases with new primers and propellant using either new bullets or homemade bullets) are practiced by many firearms enthusiasts, especially in the United States. Reloading can substantially reduce the cost of ammunition, provide ammunition for firearms chambered for unusual or obsolete cartridges, and give scope for "tailor-made" designs. Both hand loading and reloading use commercially available primers and propellants which are basically the same as those used in commercially manufactured ammunition.

Reloaders use either commercially available bullets, or make their own from commercially available lead alloy bars or a range of suitable scrap metal alloys. Two specifications for manufactured bullets for reloading purposes are:

1. 90% lead, 6% antimony, and 4% tin
2. 83.5% lead, 11.5% antimony, and 5% tin

By using varying amounts of tin and/or antimony to harden lead, reloaders can control the hardness of their bullets. They can also cast composite bullets, one mold casting the nose portion and another mold casting the base portion. The two portions are then glued together using epoxy glue.

Sources of scrap metal suitable for use in bullet making include old lead pipes, old cable sheathing, lead sheeting from old roofs, commercial lead wire, and scrap "tin" such as pewter, high-speed bearings, 50/50 bar solder (lead and tin), and basic white metal (92% tin and 8% antimony). The main sources for reloaders are wheel weight metal (approximately 90% lead, 1% tin, and 9% antimony) and printing type metal of which there are five types ranging in composition from 62% to 94% lead, 3% to 15% tin, and 3% to 23% antimony. Linotype is the most commonly used type metal (85% lead, 4% tin, and 11% antimony).

Provided that the homemade bullets meet the criteria for hardness and accuracy the reloader is not overly concerned with the exact composition. An alloy of 90% lead, 5% tin, and 5% antimony is generally the favored mix. Cast bullets may be further hardened by heat treating. For heat treating to be effective, there must be a small amount of arsenic in the alloy.

Other Projectile Types

Less common and unusual projectiles, such as exploding bullets; saboted subcaliber bullets; flare loads; wax, rubber, plastic, and wooden bullets; frangible bullets; tear gas bullets and canisters; baton rounds; flechette cartridges; poisoned bullets; multiple loads; shot loads for pistols and revolvers; and other special purpose projectile types, are known and are occasionally encountered in forensic casework.

Exploding Bullets

Exploding bullets are not a recent concept. About the middle of the nineteenth century two types of exploding rifle bullets were invented, one using a percussion cap and black powder and the other using mercury fulminate. Figure 11.5 illustrates both types.

During World War II several types of exploding rifle bullets were developed containing an integral striking pin. One contained a firing pin inside a brass bushing. The bushing was free to slide backward and forward inside a copper cup which had a small hole in the forward end. On firing, the pin moves back against the lead core. On impact, it moves forward and the point protrudes through the hole in the forward end of the copper cup and detonates the explosive mixture. This is illustrated in Figure 11.6.

Another similar design contains a firing pin which on impact detonates 0.42 g of mercury fulminate (42%) and potassium chlorate (58%) mixture. An exploding/incendiary bullet with its own firing pin was also developed,

Projectiles

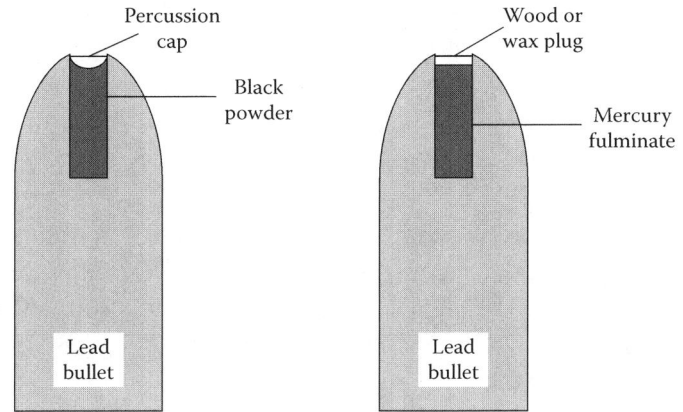

Figure 11.5 Exploding bullets.

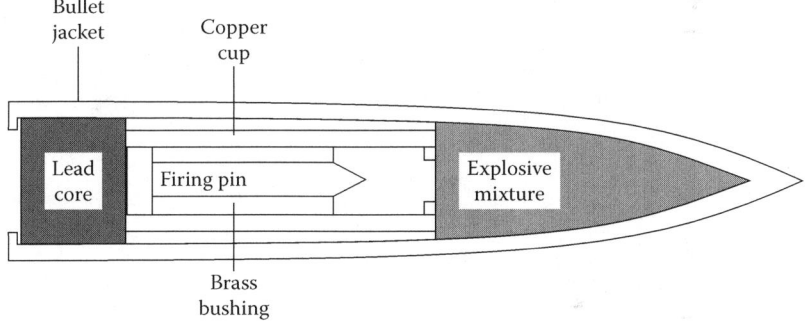

Figure 11.6 Exploding bullet.

with the incendiary component 5 grains of white phosphorus and the explosive component 7 grains of a mixture of lead styphnate, barium peroxide, and calcium silicide. This is illustrated in Figure 11.7.

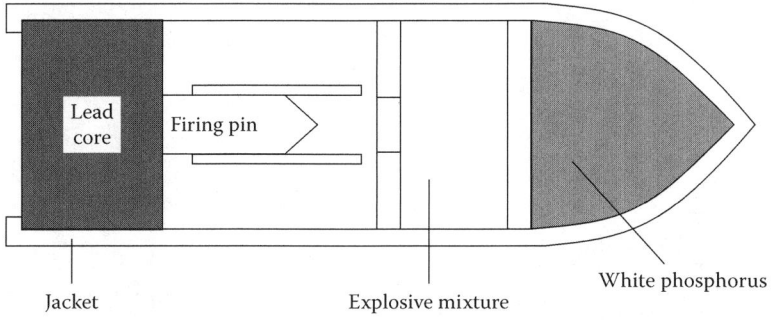

Figure 11.7 Exploding/incendiary bullet.

A further design of exploding rifle bullet, without a firing pin, contained a primer with ~40 mg mercury fulminate in the nose of the bullet which on impact ignited compressed black powder and this in turn detonated a metal cup containing an explosive mixture of potassium chlorate (56%) and antimony sulfide (44%). PETN explosive has also been used in exploding rifle bullets.

Modern explosive bullets are available in the United States from Velet Cartridge Company and Bingham Limited (Exploder/Devastator). They are offered in a range of pistol and revolver calibers. The Velet exploding bullet is a normal semijacketed hollow point bullet with black powder or Pyrodex as the explosive ingredient and with a primer cap sealing the cavity. Depending on the caliber a lead pellet may also be present in the cavity. On impact the primer cup ignites and detonates the explosive mixture. Velet also manufactured mercury-filled projectiles. The Exploder bullet is basically the same as the Velet exploding bullet except that the lead shot is omitted.[75]

The assassination attempt on President Reagan involved the use of Devastator exploding bullets. These consisted of standard 0.22" LR caliber, copper-coated lead hollow point bullets, modified by deep drilling to accept a tiny aluminum alloy canister containing about 24 mg of lead azide. Nitrocellulose lacquer was used to seal the base of the canister. RDX explosive was originally used in these loads but was replaced by lead azide.[76,77]

Flare Loads

These loads contain some sort of pyrotechnic composition and can be used for signaling purposes, to provide a source of light for brief periods, or as fireworks. They are mainly available in 38" Special and 9 mm Parabellum calibers or in 12- and 16-gauge shotgun cartridges.[78,79]

Wax Bullets

Wax bullets are used for mock dueling and fast-draw contests, short-range practice (training), and in blank ammunition. Rubber and plastic bullets are also used for short-range practice. Such ammunition is usually powered by a pistol primer.

Wood Bullets

Wood bullets are often used in blank ammunition where automatic weapon functioning is required. They can be lethal at short range if some form of bullet breakup device is not fitted to the muzzle of the firearm. Some wood bullets may be sabots for subcaliber projectiles.

Projectiles

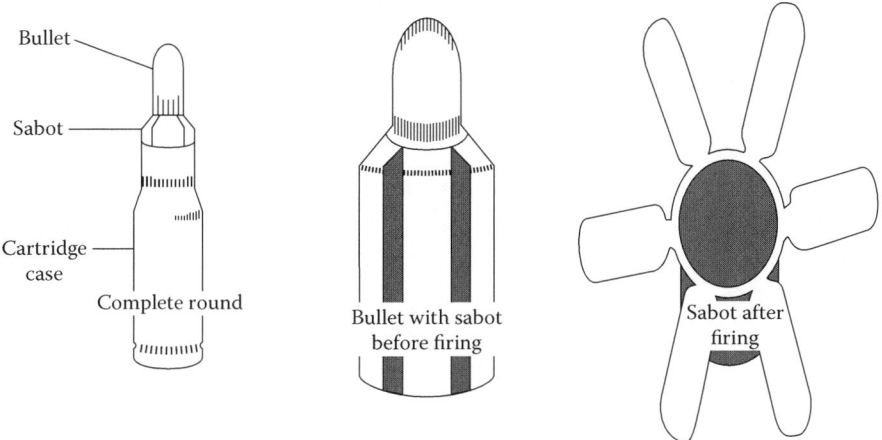

Figure 11.8 Remington "Accelerator" round.

Saboted Subcaliber Bullets

A sabot (discarding sabot) is a lightweight plastic container that encloses the lower portion of a bullet. The sabot containing the bullet is seated into the neck of the cartridge case in much the same way as a conventional bullet. The sabot is used to produce very high bullet velocities when a smaller caliber bullet (smaller weight) is fired from a larger caliber barrel. Very high velocity results from firing a lighter bullet in a larger caliber gun, the sabot forming a sort of bushing between the barrel and the bullet and acting as a gas seal and a pusher plug. The Remington "Accelerator" round is an example of this style of ammunition and is illustrated in Figure 11.8.

As the sabot and bullet move down the barrel together they both start to spin from the twist of the rifling. By the time the bullet and sabot leave the muzzle the spin rate can be in the region of 3,000 revolutions per second. The substantial centrifugal force generated opens the front end of the sabot. The sabot, as it is light weight and blunt, separates from the bullet by air resistance about 18 inches from the muzzle. Sabots can travel up to 100 yards and because of its speed of travel it is a dangerous projectile.

The forensic implications of discarding sabots are the fact that the bullet will bear no rifling marks from the barrel of the weapon and one weapon could be used to fire a number of different subcaliber projectiles.[80,81]

Multiple Loads

Ammunition has been manufactured with more than one bullet loaded inside the cartridge case. The idea is to increase the "stopping power" of the ammunition. Up to four bullets loaded in tandem within a single cartridge case are known, with the bullets reduced in size and weight compared to a normal bullet loading. Such rounds of ammunition are rare. If a person were shot at close range with one round of such ammunition there would be a single entrance wound and more than one bullet could be recovered from the body.[82,83] At greater range there could be more than one entrance hole from a single shot. A variation on this theme is stacking projectiles one over the other and then enclosing them in a normal gilding metal jacket.[84] Of historical interest is the 7.62 mm Nagant revolver ammunition. The bullet is contained completely within the cartridge case, the neck of which is coned inward in front of the bullet.[85]

Special Purpose Ammunition Types

The Glaser safety slug is a high velocity projectile with a copper jacket filled with very small lead shot pellets and sealed at the tip with a frangible plug. On striking the target the pressure on the tip prevents it from fragmenting, but on penetrating the target the plug fragments causing the bullet to disintegrate, thereby releasing its shot charge. This substantially increases its stopping power by delivering all of its energy to the target, preventing overpenetration and substantially reducing ricochet hazard.

A bullet called Hydra-Shok is designed for increased expansion within the body (mushrooming) and thus causing greater tissue disruption than with conventional ammunition. The bullet is a hollow point with a central post. The central post diverts hydrostatic pressure as the bullet passes through soft tissue thereby assisting the mushrooming of the bullet.[86] This is illustrated in Figure 11.9.

Another bullet design to assist bullet expansion is the insertion of a steel or lead ball into the cavity of a hollow point bullet, which on contact with the target is driven back into the cavity, thereby aiding expansion.

Ultra-Shock ammunition is a combination of a central post (a stainless steel screw) with a lead shot pellet over the head of the screw.

An unusual bullet design is the PMC Ultra-Mag ammunition. The bullet is machined from solid bronze (a range of copper alloys, usually with tin but sometimes with other additives such as P, Min, Al, Si, Zm) and is a hollow tube with a plastic plug at the base to form a gas seal. It is essentially a tube

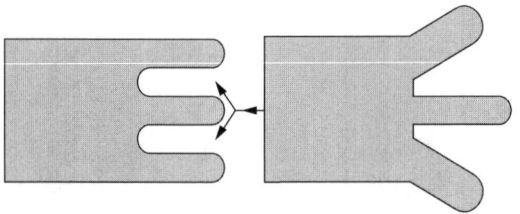

Figure 11.9 Hydra-Shok bullet.

traveling at high velocity and it slices a tubular section completely out of the center of the wound channel, thereby creating a very destructive, permanent wound.[87]

An aluminum bullet with a wounding capacity up to ten times greater than a conventional lead bullet has been developed for dealing with hijackers and hostage takers. It is claimed that it incapacitates quicker than ordinary bullets because of its lighter weight and consequent higher velocity. On hitting the target it reduces speed suddenly because of its light weight and does not pass through the body, thereby transferring all its kinetic energy to the target. The bullet has a self-lubricating nylon coating which is made to bond to the aluminum by first giving the metal some form of chemical coating.[88]

Another round of ammunition developed specifically for law enforcement use is the BAT (Blitz-Action-Trauma) ammunition. It is designed for high "stopping power" and for the projectile to have a short range of travel. The bullet is unjacketed and is made from solid copper alloy with a differing diameter hole all the way through the bullet from nose to base. This is illustrated in Figure 11.10.

A plastic plug fills the larger hole and part of the smaller one. On firing the cap separates from the bullet due to gas pressure acting through the hole in the base. Because of the lighter weight of the bullet compared to a conventional bullet it loses velocity, and hence energy, much faster, and consequently is less of a risk to innocent persons should the bullet miss its target.

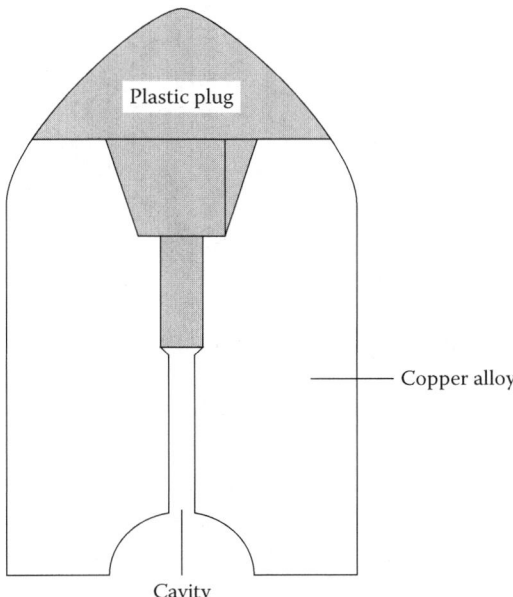

Figure 11.10 BAT bullet.

Because of the larger nose cavity the bullet deforms rapidly in the target, substantially diminishing the risk of overpenetration or ricochet, both of which could injure innocent bystanders.[89]

A bullet specifically designed to shatter car windshields and to retain sufficient energy to wound the occupant is manufactured under the name of Equaloy. It is designed not to exit from soft targets. The bullet had a semi-wadcutter profile, an aluminum alloy core, and is coated with white nylon.[90]

MagSafe bullets are bullets with gilding metal jackets, filled with an epoxy resin which has number 6 size hardened lead shot pellets embedded in it. Because of its substantially reduced weight, very high velocities are achieved. On impact the epoxy resin breaks up and releases the shot pellets, thereby transferring all the kinetic energy to the target.[91] It also substantially reduces overpenetration and ricochet hazards.

Splat multipurpose ammunition (synthetic plastic loaded ammunition for training) uses a metal-filled plastic material which is injection-molded into the shape of a bullet. The bullets are much lighter than conventional equivalents and consequently achieve much higher velocities. They are frangible on impact and virtually eliminate overpenetration and ricochet hazards. Splat ammunition is available in a range of calibers and the mass, velocity, and frangibility of the bullets can be varied to suit the intended purpose. The .38" Special caliber aircraft load is designed to safeguard against perforation of aircraft windows and body panels while engaging hijackers and other rounds are designed to penetrate car bodywork and injure the occupants. A 12-gauge shotgun slug is designed for attacking door locks and hinges to gain entry in hostage situations and on doing so it completely disintegrates. Splat bullets produce wound cavities up to ten times larger than conventional lead bullets, do not over penetrate, and all the energy is transferred to the target.[92]

Shell X-Ploder bullets are self-defense short range bullets with a compressed core of shot. Upon impact they come apart causing massive tissue destruction. They are nonricocheting.

Poisoned Bullets

The concept of a poison contained within a bullet is not new and has been experimented with over a long period of time. As all poisons, in the quantities capable of being administered via a bullet, take some time to kill, the tactical use of such ammunition is of limited military or civilian use. However, it is an attractive concept for an assassin. If the bullet itself fails to kill, the backup system of the poison offers a twofold method of attack.

In 1892, Lagard transmitted anthrax to animals by shooting them through their soft tissue with infected bullets.

A poisoned bullet containing about 38 mg of aconitin (a vegetable alkaloid; 4 mg is a lethal dose) was manufactured in the former Soviet Union.

Projectiles

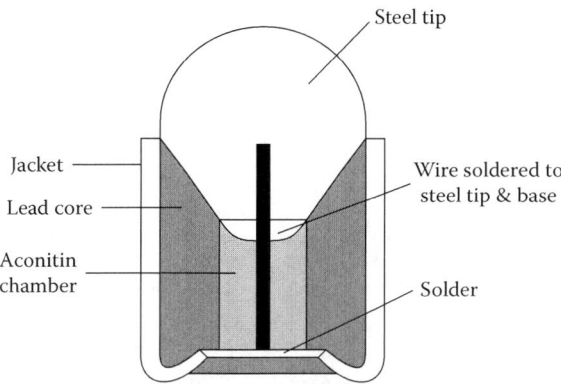

Figure 11.11 Soviet designed poisoned bullet.

Aconitin is prepared from the roots of a flowering plant known as *Aconitum napellus,* three species of which grow wild in the United States and at least one in the Soviet Union. For centuries aconitin has been used to tip poisoned arrows. Figure 11.11 illustrates the Soviet design. On impact the steel wedge is driven backward, expanding the slotted jacket and lead core. The bullet fragments within the target releasing the poison.

A German modification of the Soviet design replaced the aconitin with a glass ampoule containing an aqueous solution of hydrogen cyanide, which is faster acting than aconitin. On impact the steel striker is driven into the bullet, fragmenting the bullet and releasing the poison. Figure 11.12 illustrates the German design.

Another German design was a jacketed bullet with a hollow in the nose. On impact the steel plug squashed the glass ampoule against the lead core, causing the fluid to pour out of the nose cavity. This is illustrated in Figure 11.13.

In August 1978, a Bulgarian defector living in Paris was shot with a tiny pellet made of 90% platinum–iridium alloy which was only 1.7 mm in

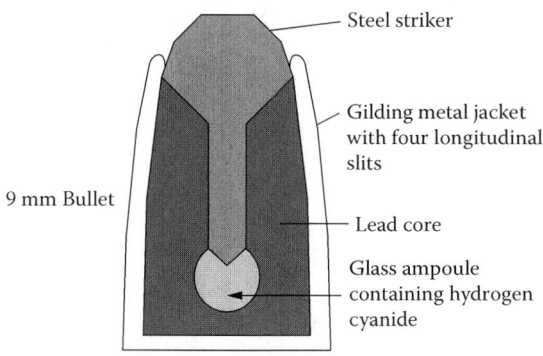

Figure 11.12 German designed poisoned bullet.

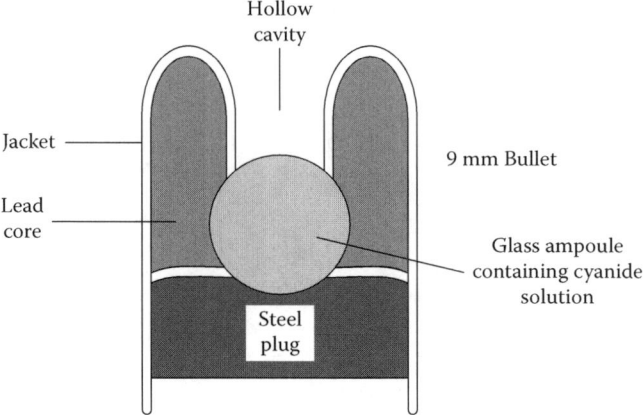

Figure 11.13 German designed poisoned bullet.

diameter. Two holes drilled in the sphere contained approximately 0.4 mg of ricin (a toxic protein derived from castor beans). This assassination attempt was unsuccessful. However, the same device was used successfully three weeks later to kill Georgi Markov in London. In this case the victim was stabbed in the thigh by an "umbrella" and injected with the pellet.

In 1975 a pistol was developed which was capable of accurately discharging a poisoned flechette over a range of 250 feet. Other dart launchers disguised as walking canes and umbrellas are also known. The flechette was coated with the shellfish poison saxitoxin. The launchers make very little noise and the toxin leaves no trace and has no known antidote. Another variation on this theme is a false 7.62 mm caliber rifle cartridge with a "separable" bullet, with the cartridge case filled with anthrax spores or botulinal toxin and designed for attacks on people within confined spaces, by breaking on the walls or floor and releasing the "poison."[93,94]

A 7.9 mm caliber false cartridge containing poison gas is also known and is designed for the same purpose. A .32" caliber poison gas cartridge designed to be fired from a pistol has also been manufactured.

Shot Loads

Pistol and revolver cartridges can be loaded with lead shot pellets, primarily for short range vermin control. The shot can be totally enclosed within the cartridge case and sealed with a wad, or the shot may be contained in a paper, cardboard, plastic, or wood sabot. A shot load/bullet combination is also manufactured in handgun calibers.[95]

Baton Rounds

As the name implies, such rounds of ammunition incorporate projectiles that are designed to incapacitate rather than kill, and are used for riot control purposes. Either plastic or rubber bullets are used, and are usually very much larger than conventional bullets. Such bullets can be fatal if the weapon is not used in the correct manner. Much smaller plastic bullets, 5.56 mm caliber, have been used for riot control by the Israeli Army and fatalities have resulted from their use.

Frangible Bullets

Until recently these were usually confined to .22" caliber and used in shooting galleries. They are designed to disintegrate on hitting the target and consequently will not ricochet. They consist of particles of iron or lead bonded in some synthetic material. One manufacturer describes a .22" short bullet as being lead or iron bonded with lanolin.[96] Such bullets are also used for training purposes.

Frangible bullets have recently become popular in a range of calibers due to the trend toward lead-free ammunition for indoor training purposes, the conventional lead bullet being replaced by a frangible nontoxic alternative coupled with the use of lead-free primers. The PMC ammunition range includes frangible bullets (93% copper and 7% polymer) coupled with a primer composition that on discharge leaves only a small trace of potash (K_2O) in the firearm, which is harmless.

Flechette Cartridge

These cartridges consist of small steel darts loaded into shot shells and rifle cartridges and are used by the military. They contain either single or multiple loads. A flechette with a sabot (Steyr Acr) has been developed and is claimed to have good penetrative ability and formidable wounding characteristics.

"Dardick Trounds"

These cartridges used a triangular plastic case with a primer in one end and the projectile in the other. They were developed for various civil and military applications but are no longer manufactured.

Tear Gas Bullets

Tear gas bullets are known to have been manufactured in a variety of calibers over the years. Israel currently makes a 7.62 mm NATO caliber tear gas bullet.[97] Tear gas guns which fire large canisters of tear gas are used for riot control and hostage situations. Smoke-producing canisters are also used for similar purposes.

References

63. "The Identification of Small Arms Ammunition," Department of Chemistry and Metallurgy, Royal Military College of Science. Technical report AC/R/31 (August 1979): 2-C-1, 2-C-2.
64. "The Identification of Small Arms Ammunition," 2-C-5.
65. Finn Aagaard, "Trophy Bonded Bullets," *American Rifleman Magazine* (May 1987): 34.
66. Ross Seyfried, "The 'Inside' Story—Big Game Bullets," *Guns and Ammo Magazine* (March 1987): 62.
67. Private communication, 1981.
68. P. Labbett, "Cartridge Corner," *Guns Review Magazine* (December 1985): 899.
69. Private communication, 1978.
70. Daniel W. Kent, *German 7.9 mm Military Ammunition 1888 to 1945* (Ann Arbor, MI: Author, 1972).
71. Private communication, 1981.
72. *Guns Review Magazine* (February 1988): 113.
73. *Guns Review Magazine* (February 1989): 113.
74. Robert Stack, "Shotgunning," *Gun World Magazine* (January 1989): 16.
75. "The Identification of Small Arms Ammunition."
76. L. G. Tate, V. J. M. DiMaio, and J. H. Davis, "Rebirth of Exploding Ammunition—A Report of Six Human Fatalities," *Journal of Forensic Sciences* 26, no. 4 (October 1981): 636.
77. "Explosive Bullets: A New Hazard for Doctors," *British Medical Journal* 284 (March 1982).
78. "The Identification of Small Arms Ammunition."
79. *Dynamit Nobel Ammunition Catalogue* (1989).
80. "The Identification of Small Arms Ammunition."
81. Bob Forker, "Swifter Than the .220 Swift," *Guns and Ammo Magazine* (February 1986): 28.
82. Vincent J. M. Di Maio, *Gunshot Wounds* (New York: Elsevier Science).
83. Wiley Clapp, "Handloading the .357 Quadraximum," *Gun World Magazine* (November 1986): 32.
84. Dave Corbin, "Monolithic Bullets," *Handloader Magazine* 125 (January/February 1987): 34.
85. P. Labbett, "Cartridge Corner," 419.
86. K. Sperry, and E. S. Sweeney, "Terminal Ballistic Characteristics of Hydra-Shok Ammunition: A Description of Three Cases," *Journal of Forensic Sciences* 33, no. 1 (January 1988): 42.
87. Jan Libourel, "PMC Ultra-Mag Ammo," *Guns and Ammo Magazine* (February 1987).

Projectiles

88. Michael Imeson, "Deadlier Than Lead," *Sunday Times Newspaper* (February 23, 1986).
89. "BAT Safety Ammo," *Guns and Ammo Magazine* (May 1983): 40.
90. Labbett, "Cartridge Corner," 859.
91. *Gun World Magazine* (April 1988): 43.
92. Nick Steadman, "Armsflash," *Target Gun Magazine* (July 1987): 57.
93. Private communication, 1979.
94. Kent, *German 7.9 Military Ammunition 1888 to 1945*.
95. Dean A. Grennell, "Reload Clinic," *Gun World Magazine* (July 1987): 12.
96. Private communication, 1979.
97. Labbett, "Cartridge Corner," 594.

Complementary Ammunition Components   12

Also present in ammunition is a range of lubricants (greases and waxes) usually confined to unjacketed bullets, as well as various sealers, varnishes, and lacquers (often colored) that are used for several purposes. These include prevention of the ingress of moisture or oil, anticorrosion measures, to act as a visual aid in the manufacturing process, to provide additional resistance of the bullet to the pressure generated by the propellant gases, and color coding to identify the type of ammunition. The bullet tip may also be color-coded to identify the bullet type. A variety of sealing discs may also be present in different types of primers, some of which contain tin.

Bullet lubricants frequently contain mixtures of some of the following: beeswax, petroleum jelly, sheep tallow, carnauba wax, molybdenum disulfide, lithium base grease, ceresin wax, powdered graphite, paraffin wax, Alox compound, and Zokorite. The mixture may be dissolved or suspended in a fast-drying solvent or applied directly while hot.

Colored lacquers and varnishes may incorporate metallic containing pigments.

Caseless Ammunition 13

Current trends in firearms and ammunition are toward lighter and easier to carry firearms and ammunition. These include the use of various plastics in the manufacture of firearms and some large, well-known manufacturers are experimenting with the development of caseless ammunition (combustible cartridge cases and primers) which, if viable, will involve major design changes in the mechanism of firearms.

The disadvantages of caseless ammunition include the very high cost of totally new weapons and ammunition manufacturing systems, relative fragility and sensitivity of the ammunition, and difficulties with weapon maintenance and repair in the field.

Advantages of caseless ammunition include space and weight saving, a cost saving over brass-cased ammunition, a material saving (copper is a critical material during wartime), and a higher cyclic rate of fire made possible because the extraction and ejection cycle of conventional firearms is no longer necessary (about 2,200 rounds per minute is possible). The higher rate of fire improves the chance of hitting the target on firing a three-round burst.

The concept of caseless ammunition is not a recent one as such ammunition, for use in breech-loading firearms, has been in circulation for well over 100 years. The concept was experimented with by different countries at various times, but with only limited success. Typical compositions consisted of nitrocellulose (12.6% nitrogen) 65%, kraft paper 15%, resin 20%, and diphenylamine (added) 1%. The German firm of Heckler & Koch has relatively recently solved many of the problems associated with caseless ammunition and produced a rifle, the HKGH, to fire 4.73 × 33 mm caliber caseless ammunition.

Caseless ammunition has no cartridge case to serve as a heat sink and barrier between the propellant and the hot chamber walls. The problem of "cook-off," that is, the heat rather than the firing pin causing the cartridge to discharge, has always been a major disadvantage with caseless ammunition. Heckler & Koch has solved this problem by the use of a propellant that contains no nitrocellulose but is based on materials more commonly associated with explosives than with propellants. The composition is a commercial secret but the new propellant has an ignition temperature about 100 K greater than nitrocellulose-based propellants.[98,99]

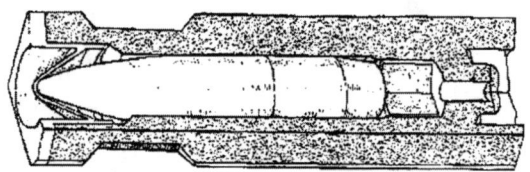

Figure 13.1 Heckler & Koch caseless ammunition.

Figure 13.1 illustrates a cross-sectional view of the 4.73 × 33 mm H&K caseless ammunition. It is rumored that Heckler & Koch have since abandoned the project.

References

98. Charles R. Fagg, "The Heckler & Koch G11," *American Rifleman Magazine* (May 1989): 30.
99. Private communication, 1991.

Blank Ammunition 14

One definition of a blank cartridge is "a percussion initiated cartridge that does not have a bullet or missile that is designed to be lethal." However, blank cartridges are dangerous if misused and serious injuries and a few fatalities have resulted from their misuse at close range. (A dummy round is not a blank as it is completely inert.)

Some blank cartridges, designed for use in cartridge tools, if discharged in a firearm may seriously damage the firearm and possibly injure the firer. Such blank cartridges generate very high gas pressure inside the cartridge tool and this is one of the reasons for the relatively heavy weight of the tool when compared to the weight of an equivalent firearm.

Blank-firing imitation/replica firearms normally have a hardened steel blockage in the barrel to prevent them from being converted to fire bulleted ammunition. A variety of blank-firing "firearms" are available commercially and blank cartridges are available in a wide range of calibers. Blank cartridges use similar priming compositions and propellants to those used in firearms ammunition and are available in rimfire and centerfire varieties. Some are powered by a primer only and some are powered by a primer plus propellant. The vast majority use brass cartridge cases but steel cases and plastic cases are also known. They can be crimp closed or closed by using a sealing wad (paper or plastic) or wax. Most, but not all, blanks have a head stamp on the cartridge case.

Blank cartridge cases have a wide range of applications:

Training (for example, weapon training, gun dogs)
Signaling (for example, starting pistol)
Cartridge tools (stud guns/power-actuated fastening tools)
Grenade launching
Antenna erecting
Engine starters
Fright guns (personal protection)
Gas guns (personal protection)
Film and theater use
"Quick-draw" contests
Humane killers (captive bolt type)
Saluting cartridges (ceremonial)

Bird scarers
Target launching
Line throwing
Mortar ignition
Balloon cable cutting
Spotting charges for practice bombs and mines
Artillery shell training adaptors
Cleaning industrial furnaces

If smoke is required the blank can be loaded with black powder. Blanks can give the flash and sound of gunfire and can even work the mechanism of a firearm if a blank firing adapter is fitted to constrict the barrel to allow the chamber pressure to be high enough, for the time period required, to operate the moving parts. This is particularly important in military training.

Some special blanks for fast-draw competitions employ a layer of slower burning rifle powder over a layer of faster burning pistol powder. The primer ignites the faster burning pistol powder which in turn ignites the slower burning rifle powder and propels it out of the muzzle of the gun. The burning powder travels far enough to burst a balloon used as the competition target. Wax bullets are also used for competitions and training where a nonlethal projectile is required.

The sealing wad or wax can be a dangerous projectile if the blank is discharged at close range. Even when no form of projectile is involved, the hot, high velocity gases emerging from the muzzle (muzzle blast) when a blank is discharged can also cause serious injury or death at close range. Exposure to these gases is used to advantage in a specialized use of live or blank cartridges, that is, a power head (bang stick or shark stick). This is a specialized "firearm" for use underwater and is meant to be discharged when in contact with the target. They are used for spear fishing, or used against sharks or alligators, for sport, self-defense, or to kill rogue animals. They normally use standard firearm ammunition and are available in a variety of handgun, rifle, and shotgun cartridges (blank cartridges can also be used). Bullets are very inefficient when used in water, traveling for only about 3 feet. By firing while in contact with the target the cartridge energy is expended directly into the flesh. The bullet has a minimal effect; the muzzle blast does the damage as the high pressure gases are forced into the flesh. Most power heads are designed to use commercial firearm ammunition and such ammunition must be waterproof. This can be achieved by coating the primer cup and case mouth with some form of varnish. For shotgun cartridges a rubber object such as a balloon can be used to seal the crimped end.

Blank cartridges can produce fatal wounds when fired at, or near, contact, and they work very well in power heads. As with firearm ammunition, blanks can also damage the eardrums when discharged.

Firearm Construction Materials 15

A range of different grades of steel are used in the manufacture of a single firearm. Chromium–molybdenum steel is the basic material for the modern firearms industry. It possesses good tensile strength, resists wear, and has good machining properties. Most .22" rimfire caliber guns, shotguns, and low pressure centerfire barrels are made from carbon steels. High-pressure centerfire barrels for sporting use are normally made from chromium–molybdenum–vanadium steel.

The grades of steel used for the many different component parts involved in a single firearm can vary from manufacturer to manufacturer, but they all contain some of the following elements in order to achieve the desired metallic properties: chromium, copper, manganese, molybdenum, nickel, phosphorus, silicon, tungsten, and vanadium. Aluminum alloys are also used in the manufacture of firearms (and telescopic sights) and contain some of the following elements: chromium, copper, iron, magnesium, manganese, nickel, silicon, and zinc.[100] Springs used in firearms may contain beryllium and copper and some parts such as sights may be attached by brazing or silver soldering. Stainless steel is being increasingly used for the manufacture of firearms. Stainless steels range from 12% to 24% chromium with other additives such as molybdenum and nickel.

The trend is toward lightweight handguns and rifles and increasing use is being made of polymers (plastics) for the manufacture of parts such as stocks, grips, frames, magazines, and so forth. Polymers have considerable advantages. They are lightweight, durable, inexpensive, noncorrosive, and easily molded to any required shape, thus eliminating the need for expensive tooling to machine a firearm to the desired shape.

A wide range of woods are used in the manufacture of solid wooden stocks for rifles and shotguns; one of the most popular is walnut. Laminated wooden stocks are also used, and consist of thin strips of wood that are impregnated with epoxy and compressed into a solid block.

Synthetic stocks are gradually replacing the traditional wooden stock due to their cheapness, ease of manufacture, lighter weight, and wear ability, and are claimed to be more stable and easier to maintain than wooden stocks. Materials such as nylon, polyurethane, fiberglass, Kevlar reinforced with fiberglass or carbon fiber, or thermoplastic resin reinforced with glass and ceramics may be used. The stocks may be hollow, or have a hollow filled with foam or a solid. Such stocks may be surface-treated with a coating such as polyurethane.

Grips may be made of wood, plastic, laminated phenolic resin, or rubber, and rubber is also used for recoil pads. Customized or ornamental firearms may have grips made from ivory; mother of pearl, or buffalo horns, and the firearm may have elaborate engraving some of which is inlaid with gold or silver. Such firearms are unlikely to be used in crime.

Homemade firearms are encountered in crime, and materials used in their construction vary markedly. Some of them are finished with household paint, and flakes of paint transferred to the criminal's clothing can be of considerable evidential value. The grips of some homemade firearms are metal and are sometimes covered with plastic adhesive tape. Plastic adhesive tape may also be used to tape magazines together for use with commercially manufactured firearms, so that the firearm can be reloaded rapidly. Fingerprints may be obtained from the tape and a comparison of tape ends with the ends of tape recovered from a suspect's home or workplace can yield very strong evidence if a physical match is obtained. The same applies to tool mark impressions from machine tools at a suspect's home or workplace.

Surface Coatings

The reasons for treating the surface of a firearm is to get rid of reflections from bright surfaces which could dazzle the firer and/or reveal the firing position to the enemy or game, to improve the appearance of the firearm, and more importantly to provide a degree of protection against corrosion.

Early firearms had no finish at all and rusted rapidly when exposed to black powder residues and atmospheric moisture. A process known as "browning" (artificial rusting) was the first attempt at a rust-resistant finish, and records on the process exist as early as 1637. The problem with browning was stopping the rusting action. The early browning finishes were known as "russetting."

"Bluing" started to replace browning by the 1800s although it originated much earlier as records exist dated 1719.

Both browning and bluing are essentially controlled artificial rusting processes using special oxidizing mixtures. The process consists of a number of stages:

1. Degreasing (by using any suitable solvent).
2. Application of browning of bluing solution.
3. Rusting (at room temperature or in a steam oven).
4. Drying.
5. Scratching (removing loose rust with a wire brush or fine steel wool).

6. Repeating stages 2 to 5 from two to six times (usually three is sufficient) until the desired color is obtained.
7. Fixing (oiling, lacquering, or waxing). Depending on the nature of the chemicals involved a further step may be necessary to neutralize or remove traces of the chemicals used before fixing.

There are hundreds of formulae for browning/bluing solutions, to the extent that it is not possible to give a typical example. However, the majority of chemicals involved are given below[101,102]:

Organic	Inorganic
Acetic acid	Ammonia, ammonium carbonate, chloride, persulfate, sulfide
Acetone	*Antimony* trichloride
Benzoin	*Arsenic*
Butyl alcohol	*Bismuth* chloride, nitrate, oxychloride
Carbon	*Boron,* i.e., boric acid
Chloroform	*Chromium,* i.e., chromic acid, oxide
Diethyl ether	*Copper* chloride, oxalate, sulfate
Ethyl alcohol	*Iodine*
Ethyl nitrate	*Iron*; ferric acetate, bisulfate, chloride, permanganate, sulfate
Formic acid	*Lead* acetate, oxide
Gallic acid	*Manganese* oxide, dioxide, peroxide, nitrate
Oxalic acid	*Mercury,* i.e., mercuric nitrate, chloride, mercurous nitrate
Picric acid	Mineral acids, i.e., hydrochloric, nitric, sulfuric
Tannic acid	*Potassium* bisulfate, chlorate, cyanide, dichromate, ferricyanide, ferritartrate, iodide, nitrate, oxalate, permanganate
Tartaric acid	*Selenium,* i.e., selenious acid
Fixing Stage	
Vaseline	*Silver* nitrate
Amber varnish	*Sodium* chloride, dichromate, hydroxide, hyposulfite, nitrate
Linseed oil	*Sulfur*
Shellac	*Tin,* i.e., Stannic chloride, oxalate, stannous chloride
Copal	*Zinc*; chloride, nitrate, sulfate
Mineral oil	Quartz sand, glass powder, water

The process of browning and bluing can be very time-consuming and labor intensive. Today, nearly all bluing (blackening) is done by the hot salt, black oxide process because of its speed and cheapness of materials. Hot salt blackening can vary in color from a blue black to a deep black depending on the concentration of the chemical solutions, the temperature, and the alloy content of the steel. The vast majority of firearms currently manufactured are finished by the use of a blackening solution containing sodium hydroxide, potassium nitrate, and sodium nitrate in the typical ratio 65:25:10, respectively.

One of the best finishes for firearm steel is "phosphatizing" (Parkerizing) but few manufacturers offer this finish other than if required for military or police markets. The process deposits a crystalline layer of phosphates on the metal surface by immersion in a bath of iron, zinc, or manganese dioxide and phosphoric acid. Of these, a manganese phosphate finish is preferred for military use.

After phosphatizing some firearms are then oil-coated to provide extra protection against corrosion. Other protective coatings applied over the phosphate coating include electrostatic spray painting, epoxy, zinc chromate, and Teflon. When the coating includes molybdenum disulfide or fluorocarbons such as Teflon, there is the added advantage of reduced friction between moving parts.

Some firearms are plated with anodized aluminum, nickel, or chromium which gives durability and good looks, and some are made from stainless steel which is much less prone to rust than conventional steel. Electroless nickel coating is an alloy coating of 88% to 96% nickel and 4% to 12% phosphorus, which is produced by chemical (not electrical) reduction of nickel on to the metal surface.

Good care and maintenance of a firearm are the best protection against corrosion. There is a wide range of commercial gun cleaning and maintenance products available.

Bore cleaning products include an electrochemical cleaning device and numerous chemical cleaning mixtures, containing both organic and inorganic compounds, the compositions of which are commercial secrets.

Lubricating products range from light mineral oil to dry lubricants incorporating molybdenum disulfide, fluorocarbons (PTFE), and other synthetic lubricants.

References

100. Private communication, 1993.
101. R. H. Angier, *Firearm Blueing and Browning* (London: Samworth Books/Arms and Armour Press/Stackpole Books, 1936).
102. C. E. Harris, "Bluing and Beyond," *American Rifleman Magazine* (December 1982): 24.

Firearm Discharge Residue IV

Firearm Discharge Residue Detection Techniques 16

Introduction

Forensic firearms casework examination encompasses the following major areas of work:

Physical	Chemical
Examination of firearms, ammunition, and associated items	Examination of swabs and clothing from suspects for firearm discharge residues
Examination of firearms and associated items for anything else of forensic interest, e.g., blood, hairs, fibers, glass, fingerprints	Examination of clothing and miscellaneous items, e.g., identification of bullet holes, differentiating between entry and exit holes, angles of fire, range of fire
Comparison macroscopy of spent bullets and spent cartridge cases	Identification of bullet strike marks, firing points, weapon hides, etc.
Identification of weapon types from spent cartridge cases and spent bullets	Chemical comparison of bullet fragments and propellants
←Serial number restoration→	
←Examination of scenes of crime→	
←Presentation of evidence in courts of law→	
←Training of laboratory staff, police, and scene-of-crime officers→	

In addition, various research and development projects are undertaken and in Northern Ireland there is also an intelligence-gathering aspect to the work.[103,104]

Firearm discharge residue work is an important aspect of the overall workload.

Firearm Discharge Residues

The gases, vapors, and particulate matter formed by the discharge of ammunition in a firearm are collectively known as firearm discharge residue (FDR) or gunshot residue (GSR) (Photograph 16.1). Anything present in the

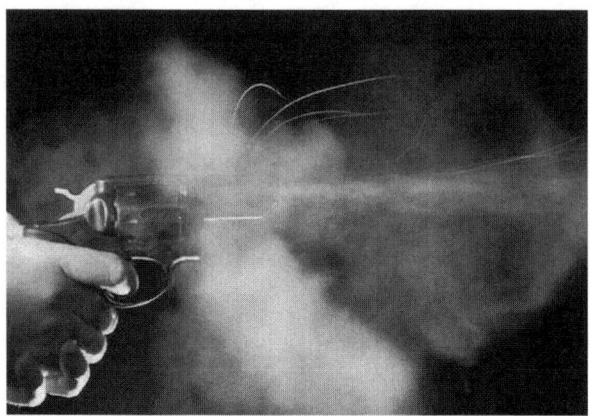

Photograph 16.1 Firearm discharge residue.

ammunition may contribute along with a possible contribution from the firearm itself. The residue consists of inorganic and organic constituents.

Inorganic constituents could originate from the primer mixture/cup/sealing disc/anvil, the cartridge case, inorganic additives to the propellant, the bullet core and jacket, metallic containing pigments in colored lacquers/sealers/lubricants, inorganic trace impurities in any component, and also from the chamber/barrel interior/muzzle of the firearm in addition to inorganic debris already present inside the firearm before discharge. The main sources of inorganic discharge residue are the bullet and the primer.

Organic constituents could originate from the primer mixture, the propellant, sealers/lacquers/lubricants from the ammunition, and also from lubricants and organic debris already present inside the firearm before discharge. The main source of organic discharge residue is the propellant.

Firearm discharge residue consists of a complex heterogeneous mixture that is claimed to be mostly particulate in nature.[105] Particulate matter can be detected on a suspect, but the possibility of vaporized/gaseous products being adsorbed on to skin or clothing surfaces also exists.

Interior ballistics has been defined as the science that investigates the way the chemical energy stored in the propellant (and to a much lesser extent the primer) is released and converted into the kinetic energy of the projectile.[106] Approximately 30% of the chemical energy is converted into kinetic energy; the rest is accommodated in the discharge residue. The discharge of a round of ammunition in a firearm produces high temperature and high pressure for a very short period of time. A typical time period from the hammer or firing pin striking the primer to the bullet or shot exiting from the muzzle is in the region of 0.03 seconds. As a result of the time period and the nature of the discharge process, only partial mixing of the constituents occurs and this accounts for the very heterogeneous nature of firearm discharge residue.

When a round of ammunition is discharged in a firearm, in addition to the projectile(s), firearm discharge residue is emitted, mainly from the muzzle but also from cylinder gaps, ejection ports, and other vents in the firearm. Some of this residue may be deposited on the skin, head hair, and clothing of the firer.

In the vicinity of the muzzle the hot propellant gases ignite and emit light on contact with the surrounding air. This effect is called muzzle flash (the bullet can also cause a flash of light whenever it strikes a target that has a hard abrasive nature[107,108]). The high pressure propellant gases at the muzzle are released into the air causing considerable turbulence and producing a powerful blast shock. This effect is known as muzzle blast.

Muzzle blast residue can be deposited on the target if the muzzle-to-target distance is less than 3 feet,[109,110] that is, a close range shooting. (The term "point blank range" is often incorrectly used in films, TV, and so forth when referring to a contact or close range shooting. The term refers to the distance a projectile will travel before dropping enough to require adjustment of the firearms sights.) Muzzle blast residue nearly always contains unburned or partially burned propellant and soot. If the firearm is fired perpendicular to the target, the resulting residue pattern will be in a roughly circular area around the entrance hole. The diameter of the circle and the density of residue depend on the distance between the muzzle and the target. The pattern size, shape, and density enable range of fire to be determined, give an indication of the angle of fire, and very occasionally the propellant granules can be identified from physical appearance as originating from a specific type of ammunition and enable a tentative identification of weapon type. Chemical analysis of the granules can yield useful information for comparison purposes with suspect ammunition. However, it must be borne in mind that a propellant can change, both in color and chemical composition, because surface coatings may be blown or burned off during discharge and the original granules may not all have a uniform composition.

High-speed photographic techniques have shown that during the discharge process some smoke emerges from the muzzle ahead of the bullet and also that just after the bullet leaves the muzzle it is surrounded by a large cloud of smoke over a short distance.[111]

The reason(s) for smoke emerging in front of the bullet could be (a) some of the discharge gases escaping in front of the bullet as the cartridge case neck begins to expand just before the bullet starts to move out of the case neck, (b) gases escaping past the bullet before the bullet completely engages the rifling, and (c) obturation not being absolutely complete, allowing gases to escape in front of the bullet through the rifling grooves.

The propellant gases expand rapidly on release into the atmosphere, accelerating to velocities much greater than that of the bullet, and this is why the bullet is surrounded by a large cloud of smoke.

As a result of these effects, FDR is deposited on the surface of the bullet. Whenever the bullet penetrates the target, due to the wiping action of the target material, some of the FDR on the bullet surface is transferred to the perimeter of the entrance hole. This occurs irrespective of the range of fire. On passing through the target, some, if not all, of the lightly adhering residue is removed from the bullet surface and consequently is not present or is present to a much lesser extent on the perimeter of the exit hole. This allows identification of bullet holes and bullet strike marks and differentiation between entrance and exit holes, which can in turn yield useful information about angle of fire, possible firing points, and relative positions of firer and target. The presence of FDR (its location, density, and nature) at a crime scene can often assist in the interpretation of the sequence of events.

Detailed analysis of the perimeter residue often yields useful information, for example, whether the bullet is unjacketed or jacketed and the nature of the jacket material. Primer type may occasionally be inferred and the presence of strontium or magnesium indicates a tracer or incendiary bullet, respectively.

Crimes involving the use of firearms are particularly serious and demand the fullest investigative effort. Thus it can be seen that the detection, identification, and quantification of FDR provide significant evidence in several areas associated with incidents involving the use of firearms.

One of the most important, difficult, and challenging aspects of the work is to connect a suspect to a firearm, or to an involvement with firearms, by the presence of FDR on the suspect's skin, head hair, or clothing, or inside the suspect's dwelling or motor vehicle.

A satisfactory test for the identification of FDR on a suspected firer has been sought by forensic scientists for many years. A satisfactory test would be one that is simple, reliable, fast, inexpensive, and conclusive. Until recently efforts have concentrated mainly on the detection of inorganic components of FDR and encompass qualitative and quantitative methods, culminating in the particle analysis method, which is the most informative method currently available. A brief outline of the most important developments follows.

Development of Firearm Discharge Residue Detection Techniques

Paraffin Test

The detection of nitrates and nitrites present in FDR using a color reaction with diphenylamine in sulfuric acid was first applied to firearms related examinations about 1911, and in 1914 Doctor Iturrioz used paraffin wax as a lifting medium for propellant residue on clothing prior to treatment with

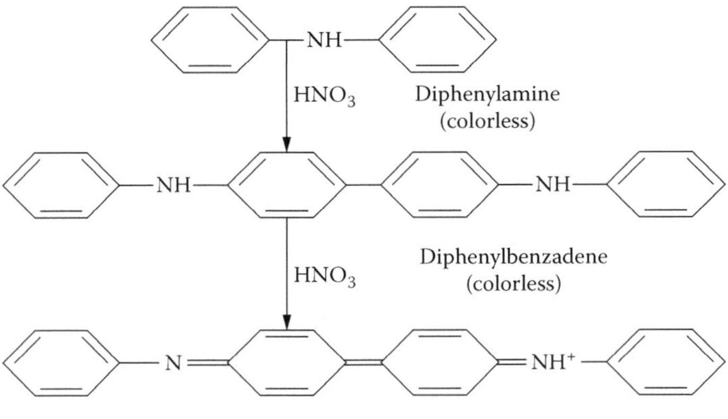

Figure 16.1 Quinoil immonium ion (blue).

diphenylamine/sulfuric acid reagent. The use of paraffin wax as a lifting medium led to the test being popularly referred to as the paraffin test.

In 1922, F. Benitez recorded this technique as a method for revealing the presence of propellant particles on the hands of a firer.[112–114] In Mexico in 1931, T. Gonzales performed a modified version of Iturrioz's test using molten paraffin wax on the hands of a firer and in 1933 demonstrated the test in the United States. The test has also been referred to as the dermal nitrate test, the diphenylamine test, and the Gonzales test.[115,116]

The test is conducted by applying melted paraffin wax to the back of a suspect's hands. With a brush, the back of the hand is coated with paraffin wax which on cooling solidifies and can be peeled off the hand. The surface of the cast that has been in contact with the skin is treated with diphenylamine/sulfuric acid reagent by dropwise addition or spraying. The reagent produces a blue color with individual particles of nitrates and nitrites. The reaction sequence is as shown in Figure 16.1.[117] The detection of dark blue spots was considered as indicative of the presence of nitrates and/or nitrates from FDR.

In 1935, the American Federal Bureau of Investigation pointed out that the test was not specific and had reservations about its use.[118] Other evaluations of the technique proved it to be completely unreliable as an indicator of FDR. Common substances such as tobacco, tobacco ash, fertilizers, certain pharmaceuticals, certain paints, and urine also give positive results.[119,120] In addition a number of oxidizing agents such as chlorates, bromates, iodates, permanganates, chromates, vanadates, molybdates, antimony$^{(V)}$, and ferric salts also give a reaction.[121,122] At the Interpol Conference in Paris in 1968 it was officially concluded that the paraffin test should no longer be used.[123]

Harrison and Gilroy Method

In 1959, Harrison and Gilroy introduced a method based on the detection of the metal-containing components of FDR.[124] The metallic components involved, namely, lead, antimony, and barium, originate from the primer and the bullet (lead and antimony). The method is based on colorimetric spot tests and involves swabbing the suspect's hands with cotton cloth damped with 0.1 M hydrochloric acid. The swab is allowed to dry and is then tested with one or two drops of a 10% alcohol solution of triphenylmethylarsonium iodide. The appearance of an orange ring indicates the presence of antimony.

The swab is then dried again and treated with two drops of freshly prepared 5% sodium rhodizonate solution to the center of the orange ring. The development of a red color indicates the presence of lead and/or barium. The swab is then dried a third time in the absence of strong light, and one or two drops of 1:20 hydrochloric acid are added to the red colored area. A blue color developed inside the orange ring is confirmation of the presence of lead. A red color remaining in the center confirms the presence of barium.

These tests were considered to be inconclusive, and the sensitivities of the colorimetric reagents used were not adequate to reliably detect the low concentrations found in actual firings.[125–129]

Neutron Activation Analysis

Neutron activation analysis (NAA) is one of the most sensitive analytical techniques for many elements. A major breakthrough came in 1964 when NAA was applied to the quantitative detection of antimony and barium in FDR.[130] (Antimony is the most valuable elemental indicator for FDR, because lead and barium are more common in occupational and environmental surveys.)

The method is based on the fact that when a sample is irradiated in a nuclear reactor for a specific length of time, atoms of some elements absorb neutrons. Nuclei that acquire an excess electron have a large excess of energy that is often released in the form of gamma rays. Nuclei with added neutrons are called radionuclides.

On placing the irradiated samples into a radio counter system capable of detecting and recording specific radiations it is possible to identify and quantify the elements of interest. If the elements emit gamma rays, the energy of the emissions and the decay lifetimes provide qualitative identification of the elements. The number of gamma rays per unit time versus the energy of the gamma rays is directly proportional to the amount of the element in the sample.[131–133]

NAA is an excellent analytical tool which has been used successfully for the detection of barium and antimony in FDR. It has been applied to the detection of FDR on suspects, the identification of bullet holes in a variety of

Firearm Discharge Residue Detection Techniques

target materials, and range of fire estimations. Copper and mercury in FDR have also been determined.[134]

NAA suffers from several major disadvantages for routine operation by most forensic laboratories:

1. Availability of and access to a nuclear reactor;
2. High equipment costs and lack of trained staff;
3. Slow throughput of samples due to time required for irradiation, cooling, and radiochemical separation;
4. Poor detection limit for lead (~10 µg), a very important element in FDR work.

Much development work has been done on the NAA technique for FDR detection, but the inherent disadvantages of the technique led to a search for other more suitable quantitative methods. Despite its disadvantages, NAA did much to increase our knowledge of quantitative aspects of the deposition and subsequent behavior of FDR.

Many alternative techniques, both qualitative and quantitative, have been investigated either for screening purposes or as primary methods. Such techniques include atomic absorption spectrophotometry, molecular luminescence, electron spin resonance spectrometry, X-ray analysis methods, and electro analytical methods. Flameless atomic absorption spectrophotometry (FAAS) is the technique that has almost completely replaced NAA.

Flameless Atomic Absorption Spectrophotometry

The FAAS method offers similar detection limits to NAA and is suitable for the determination of low levels of lead. Equipment costs are reasonable and the instrumentation is commonplace in many analytical laboratories. A large number of metallic elements, over a wide concentration range, extending down to ultra-trace level, can be analyzed, thus making the technique versatile and useful for other forensic applications as well as FDR detection. Apart from cost, the main advantages are simplicity, speed of analysis, and "in house" operation. One disadvantage of FAAS is that it is not capable of simultaneous multielement analysis.

In FAAS the sample is electrically heated to a high temperature, thus breaking the chemical bonds and enabling individual atoms to float freely in the sample area. These ground state atoms are then capable of absorbing ultraviolet or visible radiation. The wavelength bands that each specific element can absorb are very narrow and different for every element. The desired element can be considered able to absorb only "resonance lines" whose wavelengths correspond to transitions from its ground state to some higher energy level.

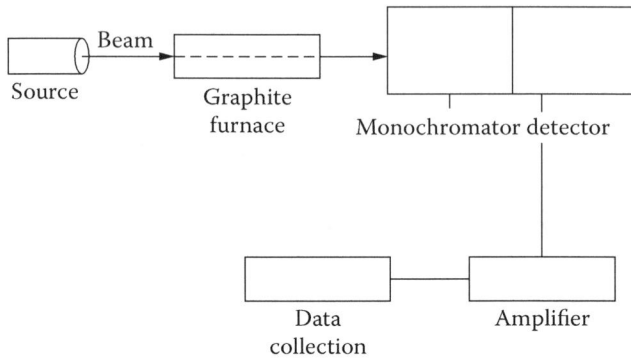

Figure 16.2 Basic components of a flameless atomic absorption spectrophotometer.

A basic atomic absorption instrument includes a source of radiation, a system for placing sample atoms in the ground state, a monochromator to separate the resonance line of interest, and finally a detector to measure the decrease in signal from the source when absorbing atoms are placed in the light beam. The magnitude of the decrease in signal is dependent on the amount of the element of interest in the sample. Figure 16.2 illustrates the components of a basic flameless atomic absorption spectrophotometer.

FAAS is the most popular technique for the quantitative determination of elements associated with FDR: lead, antimony, barium, copper, and mercury. Other relevant elements have also been determined, and the use of FAAS for FDR detection is well documented in the literature.[135–138]

All bulk elemental analysis methods, such as NAA and FAAS, suffer from the serious disadvantage of lack of specificity in that the elements detected are not unique to FDR but also occur from occupational and environmental sources. Many surveys were carried out to determine background levels of lead, antimony, and barium on the hands of people not involved with firearms. Some surveys also included copper and mercury. Both general and occupational data were gathered and threshold levels established for each of the elements. The threshold level may be defined as the level above which the results may be significant and correlate to the discharge of a firearm. The best that could be stated was that the levels detected were consistent with the discharge of a firearm but could not be taken as conclusive proof of the presence of FDR.

A more definite method was sought resulting in the particle analysis method, which is claimed to conclusively identify FDR particles.[139–144]

Particle Analysis Method

This method employs a scanning electron microscope equipped with elemental analysis capability (SEM/EDX) and combines details of the morphology and elemental composition of individual FDR particles. A particle classification scheme was developed and is based on the elemental composition and morphology of individual FDR particles and is used to classify particles as one of the following: (a) of nonfirearm origin, (b) consistent with originating from the discharge of a firearm, or (c) definitely from the discharge of a firearm. The ability to identify FDR particles uniquely and to distinguish them from environmental sources of lead, antimony, and barium eliminates the threshold problem inherent in bulk elemental analysis.

In the SEM a beam of electrons is accelerated from a hot tungsten filament down a column by a high anode potential (up to 50 kV), and by the use of three electromagnets (electromagnetic lenses) it is focused on the surface of the sample (specimen) which is mounted on a metal stub. The primary electron beam interacts with the elements at the surface of the sample causing, among other things, secondary electrons, backscattered electrons, and characteristic X-rays to be emitted from the sample surface. Figure 16.3 illustrates the different interactions.

These three effects are utilized in the particle analysis method as the surface of the sample is traversed (scanned) by the electron beam. The secondary electrons are used to view the sample, the backscattered electrons are used to identify likely FDR particles, and the X-rays are used to provide details of the elemental composition of the particles.

Scanning is achieved by the use of scanning (raster) coils which cause the primary beam to be electromagnetically deflected across a given area of the sample surface. The raster pattern of the beam is synchronized with the scanning pattern of the cathode ray tube.

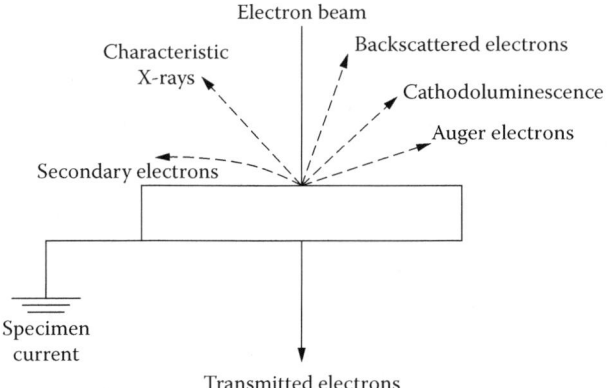

Figure 16.3 Electronic interaction with sample surface.

The low energy secondary electrons (less than 50 eV) ejected from the sample surface are attracted to a scintillator on the end of a perspex light guide, the other end of which is in contact with the window of a photomultiplier tube. The amplified signal is displayed on a cathode ray tube which records the image. The secondary electrons are attracted to the scintillator by a positively charged cage surrounding the end of the light guide. The number of electrons reaching the scintillator is dependent on (a) the topography of the sample surface since this will influence whether or not a particular area on the sample surface is visible to the primary beam and detector and (b) the elemental nature of the sample surface as this will affect the energy of the secondary electrons and consequently their susceptibility to the cage potential. The topography is the more important of the two factors.

Backscattered electrons are those electrons that have undergone single or multiple scattering events, and escape back through the surface of the sample with energies greater than 50 eV. Backscattered electrons travel in straight lines and because of their higher energies they are not attracted to the secondary electron detector. Backscattering increases as the atomic number of the sample increases and there is such a strong correlation with atomic number that the relationship forms the basis for a contrast mode in SEM.

FDR particles contain elements with high atomic number (heavy metals) and this fact is used to aid the search for FDR particles located among many other particles with similar morphology but from non-firearm-related sources. The backscattered electron image is displayed on a separate screen. All particles containing heavy metals show stronger emission and appear as bright areas on the screen; only the bright particles are potential FDR and consequently only these particles need to be analyzed. Without the aid of a backscattered image, all particles with similar morphology to FDR would need to be analyzed.

Characteristic X-ray emission is one process by which an atom may stabilize itself following ionization by the electron beam. When an electron from an inner atomic shell has been dislodged, an electron from an outer shell will replace it. The difference in energy between the initial and final state may be emitted as X-radiation. The various shells of an atom have discrete amounts of energy. It follows that their energy difference, emitted as X-radiation, is also a discrete quantity and is characteristic of the atom from which it was released. X-ray spectroscopy in the SEM, as used for FDR work, involves the identification of radiation of specific energies using a special detector, and can identify elements heavier than sodium. Detection of lighter elements is not possible using a conventional detector due to the inefficiency of X-ray generation for the lighter elements. Figure 16.4 illustrates the basic components of a SEM suitable for FDR work.

Samples that do not conduct electricity and heat (insulators) cause problems in the SEM unless they can be made conductive by some means.

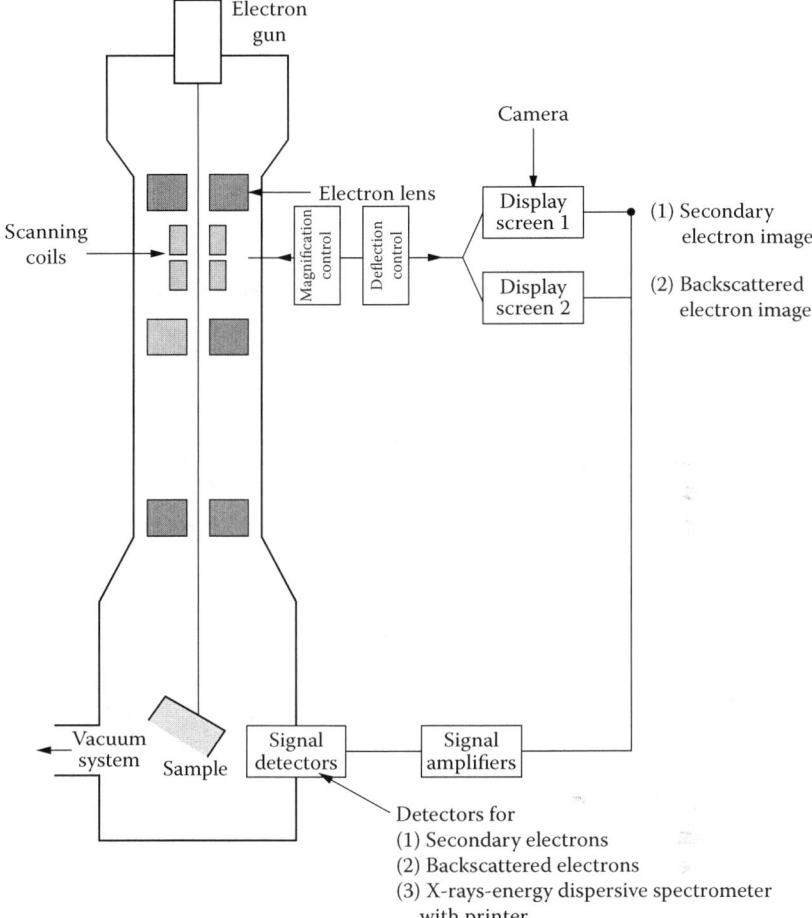

Figure 16.4 Basic components of a scanning electron microscopic suitable for FDR detection.

Problems include charging and overhearing of the sample. Coating samples with a thin layer of conductive material helps to overcome the problems.

Samples for FDR examination are taken from a suspect's skin and/or clothing surfaces by a nondestructive sampling technique and prepared for examination in the SEM. FDR samples are coated with a very thin layer of carbon using either a vacuum coater or sputter coater before introduction to the SEM. The examination involves searching for FDR particles among many other particles from occupational and environmental sources, and is a labor-intensive and time-consuming task. FDR particles are recognized by a combination of morphology and elemental composition, and are classified as FDR using a scheme developed for the purpose. The classification system will be discussed later.

Particle analysis is the most informative method to date for the identification of FDR particles. It does, however, suffer from several major disadvantages including high cost of instrumentation and lengthy and tedious procedures requiring specialized staff. Since its introduction serious attempts have been made to solve the time problem. These include the use of backscattered electron images, automation of the search procedure, and sample manipulation to pre-concentrate the sample prior to SEM examination.[145–151]

Despite all the considerable improvements, the particle analysis method remains a lengthy and costly procedure. These disadvantages have renewed interest in the possibility of detecting the organic components of FDR, either as a primary method or as a screening technique.

Detection of Organics in Firearm Discharge Residue

Chromatographic techniques are the main methods that have been used to separate, detect, and identify organic components of FDR.[152–162] Other methods considered include molecular luminescence,[163] infrared spectroscopy,[164] Raman spectroscopy,[165] electron spin response spectrometry,[166] microchemical crystal tests,[167,168] ultraviolet spectroscopy/nuclear magnetic resonance/polarography.[169]

Many of the organic constituents of FDR are explosive or explosive-related compounds and much of the work already done on the detection of explosive residues can be extended to include FDR. Explosives and their residues are usually analyzed using chromatographic techniques. Chromatography is the general name given to the methods by which two or more compounds in a mixture physically separate by distributing themselves between two phases: (a) a stationary phase, which can be a solid or a liquid supported on a solid, and (b) a mobile phase, either a gas or a liquid which flows continuously around the stationary phase. The separation of individual components results primarily from differences in their affinity for the stationary phase.

Of all the methods investigated for organics in FDR it would appear that high-performance liquid chromatography (HPLC) and gas chromatography with a mass spectrometer detector (GC/MS) are currently the most promising.

High-Performance Liquid Chromatography

In liquid chromatography the mobile phase is a liquid. The stationary phase is a solid contained inside a long narrow tube called "the column." Separation of mixtures occurs during passage through the column. The separation can occur by different mechanisms depending on the type of material in the mixture and the column, for example, adsorption, exclusion, ion exchange, partition. Whatever the mechanism, the volume of mobile phase at which a component elutes is constant and characteristic of that component, for a given chromatographic system. Thus the retention volume (time) can be used for qualitative identification. If the detector response is related to the amount of component in the sample, then the area under the chromatographic peak gives a measure of the quantity present.

The components in a mixture separate in the column and exit from the column at different times (retention times). As they exit, the detector registers the event and causes the event to be recorded as a peak on the chromatogram. A wide range of detector types are available and include ultraviolet adsorption, refractive index, thermal conductivity, flame ionization, fluorescence, electrochemical, electron capture, thermal energy analyzer, nitrogen-phosphorus. Other less common detectors include infrared, mass spectrometry, nuclear magnetic resonance, atomic absorption, plasma emission.

Chromatography is a very versatile technique offering a wide range of solid phase materials and detector types which can deal with very complex mixtures. In practice all materials and conditions used in the instrument are carefully chosen to match the type of sample mixture involved. This includes selection of stationary phase (chemical and physical properties); column type and length; sample pretreatment, operational temperatures, pressures, and flow rates; physical and chemical nature of mobile phase; detector type; and so forth. Detection to nanogram level is quite common and some systems can detect to picogram level using very small volumes of sample.

One of the major differences between normal liquid chromatography and HPLC is the column. In liquid chromatography the solid phase consists of large porous particles (75 to 200 µm) packed into columns with internal diameters of 1 to 5 cm. Very low pressures are required to permit solvent flow (mobile phase) through the large particles in the column. Flow rates are very slow and separation times are long. HPLC uses narrower column diameters and solid phase particles that are much smaller and more uniform (3 to 10 µm). This leads to much larger back pressures than with liquid chromatography and high pressure pumps are required. However, the column efficiency is increased 10- to 100-fold and separation times decreased compared to liquid chromatography. Separation methods employed in HPLC include normal phase, steric-exclusion, ion exchange, ion pair, and reversed phase.

Gas Chromatography/Mass Spectrometry

A gas chromatograph is an apparatus consisting of an injection port connected to a column that has a detector at its outlet end. The column is contained in an oven that is electrically heated, either isothermally or at a programmed rate. A stream of inert carrier gas, usually helium, is introduced into the injection port and flows through the column and detector. The injection port is a heated region that is sealed from the outside environment by a silicone rubber septum through which the sample is injected using a hypodermic syringe. Individual sample components are vaporized and travel through the column at a rate dependent on their interaction with the material used to pack the column. The detector registers and records the output from the column. In this case the detector is a mass spectrometer.

Mass spectrometers are sophisticated instruments that produce, separate, and detect positively charged gas phase ions. (Negatively charged ions can also be investigated but the abundances of negative ions are 10 to 1,000 times less than those of positive ions and negative ion mass spectrometry is less frequently used.) Pure compounds can be identified by their characteristic ions and the result is nearly always unambiguous.

The mass spectrometer consists of an inlet system, an ionization device, a mass analyzer, and an ion detector, and the system is kept at a vacuum of about 10^{-4} to 10^{-7} Torr. (There are various options available for all four basic components which illustrates the versatility of the technique.) The following describes an electron impact-quadruple mass spectrometer.

Neutral gaseous molecules entering the ionization area are bombarded with electrons to "smash" the compound of interest to yield positively charged ions. Ionization is often followed by a series of spontaneous competitive decomposition reactions (fragmentation) which produce additional ions. The instrument operates under a high vacuum to prevent absorption of the charged particles by air molecules.

The ions are then extracted from the ion source, collimated, and accelerated by an electric potential applied across a series of metal plates with exit apertures, before entering the quadrupole mass analyzer. The quadrupole mass analyzer consists of four circular parallel rods. The rods are electrically connected in diagonally opposite pairs and mass separation is achieved by applying a DC potential, positive to one pair and negative to the other and superimposing a radio frequency AC potential which differs in phase by 180° between the pairs of rods. The peak values of the AC voltage are greater than the DC voltage so that the "positive" pair is sometimes negative, and vice versa.

This produces a dynamic arrangement of electromagnetic fields resulting in certain ions taking a stable path through the analyzer and reaching the detector, whereas other ions take unstable paths and are filtered out before reaching the detector. The detector consists of an electron multiplier that magnifies the current generated by the ions striking the detector by a factor of 10^6 to 10^9. The event is recorded by some form of data system. The mass spectrum is a plot of the electron multiplier output (intensity) versus the quadrupole electrical settings and produces a characteristic pattern of fragments for the compound under investigation.

Mass spectrometry on its own is not suitable for FDR work because generally speaking "pure" compounds must be analyzed. Samples for FDR examination taken from skin and clothing surfaces are complex mixtures containing many unpredictable contaminants from occupational and

Firearm Discharge Residue Detection Techniques

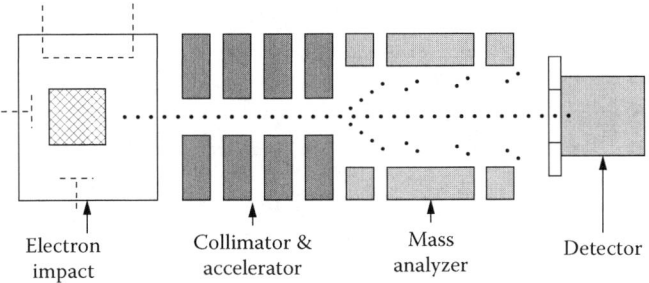

Figure 16.5 Quadrupole mass spectrometer.

environmental sources as well as skin salts and lipids. The problem is solved by separation of the constituents using gas chromatography prior to introduction to the mass spectrometer.

Combined gas chromatography–mass spectrometry (GC-MS) takes advantage of the separating power of the gas chromatograph and the identification power of mass spectrometry. The gas chromatograph separates the components and provides retention time data and the mass spectrometer identifies the components. The combined instrumentation has the potential to provide very useful information in FDR casework. Figure 16.5 illustrates a quadrupole mass spectrometer.

References

103. J. S. Wallace, "Firearms Examination in a Terrorist Situation." Paper presented at the International Congress on Techniques for Criminal Identification and Counter Terrorism, Jerusalem, Israel, February 24–28, 1985.
104. G. E. Montgomery, "The Work of the Firearms Examiner in a Terrorist Situation." Paper presented at the AFTE annual training seminar, Houston, TX, June 9–14, 1991.
105. G. M. Wolten, and R. S. Nesbitt, "On the Mechanism of Gunshot Residue Particle Formation," *Journal of Forensic Sciences* 25, no. 3 (July 1980): 533.
106. C. G. Wilber, *Ballistic Science for the Law Enforcement Officer* (Springfield, IL: Charles C Thomas, 1977), 50.
107. J. S. Wallace, "Bullet Strike Flash," *AFTE Journal* 20, no. 3 (July 1988).
108. J. S. Wallace, and K. J. Arnold, "Investigation of a Fire in an Indoor Firing Range, *AFTE Journal* 19, no. 3 (July 1987).
109. I. C. Stone, and C. S. Petty, "Examination of Gunshot Residues," *Journal of Forensic Sciences* 19 (1974): 784.
110. R. Saferstein, *Criminalistics—An Introduction to Forensic Science*, 2nd ed. (Englewood Cliffs, NJ: Prentice-Hall, 1981), 350.
111. C. L. Farrar, and D. W. Leeming, *Military Ballistics—A Basic Manual* (London: Brassey's Defence Publishers, 1983), 59.
112. I. Castellanos, "Dermo-Nitrate Test in Cuba," *Journal of Criminal Law and Criminology* 33, no. 6 (March/April 1953): 482.

113. F. Benitez, "Agunas consideraciones sobre las manchas producidas por los disparos de armas de fuege," *Revisión de Médico Legal de Cuba* 1 (1922): 30.
114. J. T. Walker, "Bullet Holes and Chemical Residues in Shooting Cases," *Journal of Criminal Law and Criminology* 31 (1940): 497.
115. I. Castellanos, "Dermo-Nitrate Test in Cuba," *Journal of Criminal Law and Criminology* 33, no. 6 (March/April 1953): 482.
116. M. E. Cowan, and P. L. Purdon, "A Study of the 'Paraffin Test,'" *Journal of Forensic Sciences* 12, no. 1 (1967): 19.
117. F. Feigl, *Spot Tests in Inorganic Analysis*, 5th ed. (New York: Elsevier, 1958), 327.
118. "The Diphenylamine Test for Gunpowder," *FBI Law Enforcement Bulletin* 4, no. 10 (1935): 5.
119. H. W. Turkel, and J. Lipman, "Unreliability of Dermal Nitrate Test for Gunpowder," *Journal Criminal Law Criminology and Police Science* 46 (1955): 281.
120. M. E. Cowan, and P. L. Purdon, "A Study of the 'Paraffin Test,'" *Journal of Forensic Sciences* 12, no. 1 (1967): 19.
121. F. Feigl, *Spot Tests in Inorganic Analysis*, 5th ed. (New York: Elsevier, 1958), 327.
122. A. I. Vogel, *Marco and Semimicro Qualitative Inorganic Analysis*, 4th ed. (London: Longman, 1954), 365.
123. A. Maehly, and L. Strömberg, *Chemical Criminalistics* (New York: Springer-Verlag, 1981), 189.
124. H. C. Harrison, and R. Gilroy, "Firearms Discharge Residues," *Journal Forensic Sciences* 4, no. 2 (1959): 184.
125. G. Price, "Firearms Discharge Residues on Hands," *Journal Forensic Science Society* 5 (1965): 199.
126. C. R. Midkiff, "Detection of Gunshot Residues: Modern Solution for an Old Problem," *Journal Police Science and Administration* 3, no. 1 (1975): 77.
127. R. R. Ruch, V. P. Guinn, and R. H. Pinker, "Detection of Gunpowder Residues by Neutron Activation Analysis," *Nuclear Science and Engineering* 20 (1964): 381.
128. M. E. Cowan, P. L. Purdon, C. M. Hoffman, R. Brunelle, S. R. Gerber, and M. Pro, "Barium and Antimony Levels on Hands—Significance as Indicator of Gunfire Residue." Paper presented at the Second International Conference on Forensic Activation Analysis, Glasgow, 1972, Paper 21.
129. M. D. Cole, N. Ross, and J. W. Thorpe, "Gunshot Residue and Bullet Wipe Detection Using a Single Lift Technique," *AFTE Journal* 24, no. 3 (July 1992): 254.
130. R. R. Ruch, V. P. Guinn, and R. H. Pinker, "Detection of Gunpowder Residue by Neutron Activation Analysis," *Nuclear Science and Engineering* 20 (1964): 381.
131. H. L. Schlesinger, H. R. Lukens, V. P. Guinn, R. P. Hackleman, and R. F. Korts, Gulf General Atomic Report No GA-9829 (1970).
132. C. R. Midkiff, "Detection of Gunshot Residues: Modern Solution for an Old Problem," *Journal Police Science and Administration* 3, no. 1 (1975): 77.
133. R. Cornelis, and J. Timperman, "Gunfiring Detection Method Based on Sb, Ba, Pb and Hg Deposits on Hands, Evaluation of the Credibility of the Test," *Medicine, Science and Law* 14, no. 12 (1974): 98.

134. M. Jauhari, T. Sing, and S. M. Chatterji, "Primer Residue Analysis of Ammunition of Indian Origin by Neutron Activation Analysis," *Forensic Science International* 19 (1982): 253.
135. A. L. Green, and J. P. Sauve, "The Analysis of Gunshot Residue by Atomic Absorption Spectrophotometry," *Atomic Absorption Newsletter* 11 (September–October 1972): 93.
136. G. D. Renshaw, "The Estimation of Lead, Antimony and Barium in Gunshot Residues by Flameless Atomic Absorption Spectrophotometry," CRE report no. 103, Aldermaston, England. Home Office Central Research Establishment (1973).
137. S. S. Krishnan, K. A. Gillespie, and E. J. Anderson, "Rapid Detection of Firearm Discharge Residues by Atomic Absorption and Neutron Activation Analysis," *Journal of Forensic Sciences* 16 (1971): 144.
138. J. S. Wallace, "Firearms Discharge Residue Detection Using Flameless Atomic Absorption Spectrophotometry," *AFTE Journal* 19, no. 3 (July 1987).
139. J. E. Wessel, P. F. Jones, Q. Y. Kwan, R. S. Nesbitt, and E. J. Rattin, "Gunshot Residue Detection," The Aerospace Corporation, El Segundo, CA. Aerospace report no. ATR-75 (7915)-1 (September 1974).
140. G. M. Wolten, R. S. Nesbitt, A. R. Calloway, G. L. Lopel, and P. F. Jones, "Final Report on Particle Analysis for Gunshot Residue Detection," The Aerospace Corporation, El Segundo, CA. Aerospace report no. ATR-77 (7915)-3 (September 1977).
141. E. Boehm, "Application of the SEM in Forensic Medicine," *Proceedings of the Fourth Annual Scanning Electron Microscopy Symposium* (1977), 553.
142. R. Diedericks, M. J. Camp, A. E. Wilimovsky, M. A. Haas, and R. F. Dragen, "Investigations into the Adaptability of Scanning Electron Microscopy and X-ray Fluorescence Spectroscopy to Firearms Related Examinations," *AFTE Journal* 6 (1974).
143. R. H. Keeley, "Some Applications of Electron Probe Instruments in Forensic Science," *Proceedings of Analytical Division Chemical Society* 13 (1976): 178.
144. J. Andrasko, and A. C. Maehly, "Detection of Gunshot Residues on Hands by Scanning Electron Microscopy," *Journal Forensic Sciences* 22 (1977): 279.
145. T. G. Kee, C. Beck, K. P. Doolan, and J. S. Wallace, "Computer Controlled SEM Micro Analysis and Particle Detection in the Northern Ireland Forensic Science Laboratory—A Preliminary Report," Home Office Internal Publication, Technical note no. Y 85 506.
146. T. G. Kee, and C. Beck, "Casework Assessment of an Automated Scanning Electron Microscope/Microanalysis System for the Detection of Firearms Discharge Particles," *Journal of the Forensic Science Society* 27 (1987): 321.
147. W. L. Tillman, "Automated Gunshot Residue Particle Search and Characterisation," *Journal of Forensic Sciences* 32, no. 1 (January 1987): 62.
148. R. S. White, and A. D. Owens, "Automation of Gunshot Residue Detection and Analysis by Scanning Electron Microscopy/Energy Dispersive X-Ray Analysis (SEM/EDX)," *Journal of Forensic Sciences* 32, no. 6 (November 1987): 1595.
149. J. S. Wallace, and R. H. Keeley, "A Method for Preparing Firearms Residue Samples for Scanning Electron Microscopy," *Scanning Electron Microscopy* 2 (1979).
150. D. C. Ward, "Gunshot Residue Collection for Scanning Electron Microscopy," *Scanning Electron Microscopy* 3 (1982): 1031.

151. A. Zeichner, H. A. Foner, M. Dvorachek, P. Bergman, and N. Levin, "Concentration Techniques for the Detection of Gunshot Residues by Scanning Electron Microscopy/Energy Dispersive X-Ray Analysis (SEM/EDX)," *Journal of Forensic Sciences* 34, no. 2 (March 1989): 312.
152. I. Jane, P. G. Brookes, J. M. F. Douse, K. A. O'Callaghan, "Detection of Gunshot Residue via Analysis of Their Organic Constituents," *Proceedings of the International Symposium on the Analysis and Detection of Explosives*, FBI Academy, Quantico (1983), 475.
153. K. Bratin, P. T. Kissinger, R. C. Briner, and C. S. Bruntlett, "Determination of Nitro Aromatic, Nitramine and Nitrate Ester Explosive Compounds in Explosive Mixtures and Gunshot Residue by Liquid Chromatography and Reductive Electrochemical Detection," *Analytica Chimica Acta* 130, no. 2 (1981): 295.
154. J. M. F. Douse, "Trace Analysis of Explosives in Handswab Extracts Using Amberlite XAD-7 Porous Polymer Beads, Silica Capillary Column Gas Chromatography with Electron Capture Detection and Thin Layer Chromatography," *Journal of Chromatography* 234 (1982): 415.
155. D. H. Fine, W. C. Yu, E. U. Goff, E. C. Bender, and D. J. Reutter, "Picogram Analyses of Explosive Residues Using the Thermal Energy Analyzer (TEA)," *Journal of Forensic Sciences* 29, no. 3 (July 1984): 732.
156. L. S. Leggett, and P. F. Lott, "Gunshot Residue Analysis via Organic Stabilizers and Nitrocellulose," *Microchemical Journal* 39 (1989): 76.
157. M. H. Mach, A. Pallos, and P. F. Jones, "Feasibility of Gunshot Residue Detection via Its Organic Constituents. Part 1. Analysis of Smokeless Powders by Combined Gas Chromatography–Chemical Ionisation Mass Spectrometry," *Journal of Forensic Sciences* 23, no. 3 (1978): 433.
158. M. H. Mach, A. Pallos, and P. F. Jones, "Feasibility of Gunshot Residue Detection via Its Organic Constituents. Part 2. A Gas Chromatography–Mass Spectrometry Method," *Journal of Forensic Sciences* 23, no. 3 (1978): 446.
159. J. B. F. Lloyd, "High-Performance Liquid Chromatography of Organic Explosives Components with Electrochemical Detection at a Pendant Mercury Drop Electrode," *Journal of Chromatography* 257 (1983): 227.
160. J. B. F. Lloyd, "Clean-up Procedures for the Examination of Swabs for Explosives Traces by High-Performance Liquid Chromatography with Electrochemical Detection at a Pendant Mercury Drop Electrode," *Journal of Chromatography* 261 (1983): 391.
161. J. M. F. Douse, "Dynamic Headspace Method for the Improved Clean-up of Gunshot Residues prior to the Detection of Nitroglycerine by Capillary Column Gas Chromatography with Thermal Energy Analysis Detection," *Journal of Chromatography* 464 (1989): 178.
162. H. R. Dales, "An Assessment of the Use of High Performance Liquid Chromatography with an Electrochemical Detector for the Analysis of Explosive Residues" (M.Sc. thesis, University of Strathclyde, August 1989).
163. J. E. Wessel, P. F. Jones, Q. Y. Kwan, R. S. Nesbitt, and E. J. Rattin, "Gunshot Residue Detection," The Aerospace Corporation, El Segundo, CA. Aerospace report no. ATR-75 (7915)-1 (September 1974), 6.
164. D. Chasan, and G. Norwitz, "Qualitative Analysis of Primers, Tracers, Igniters, Incendiaries, Boosters, and Delay Compositions on a Microscale by Use of Infrared Spectroscopy," *Microchemical Journal* 17 (1972): 31.

165. J. E. Wessel, P. F. Jones, Q. Y. Kwan, R. S. Nesbitt, and E. J. Rattin, "Gunshot Residue Detection," The Aerospace Corporation, El Segundo, CA. Aerospace report no. ATR-75 (7915)-1 (September 1974), 6.
166. L. A. Franks, and R. K. Mullen, "Time Dependent Electron Paramagnetic Resonance Characteristics of Detonated Primer Residues," U.S. Dept. of Justice, Law Enforcement Assistance Administration, National Institute of Law Enforcement and Criminal Justice (November 1972).
167. C. R. Newhouser, "Explosives Handling Detection Kit," Technical Bulletin 33-72 (Washington, DC: International Association of Chiefs of Police, 1972).
168. F. T. Sweeney, and P. W. D. Mitchell, "Aerosol Explosive Indicator Kit," Final Report F-C 377G02, LWL Task No. 24-C-74 (Philadelphia, PA: Franklin Institute Research Labs, June 1974).
169. *CRC Critical Reviews in Analytical Chemistry*, vol. 7 (Boca Raton, FL: CRC Press, 1977).

Properties of Firearm Discharge Residue   17

Formation

The development of the particle analysis method for FDR detection and identification involved consideration of how the particles are formed and of their physical and chemical nature.[170,171]

At the present time the exact mechanism(s) of formation of the particles can only be deduced from considerable practical experience and limited experimental work, most of which, if not all, has been concerned with the elemental content of FDR particles.

An extensive and very valuable study of the nature of FDR particles was conducted at the Aerospace Corporation, California, in the mid-1970s. The vast majority of this chapter is based on its findings. From firings involving a wide range of handguns and ammunition, extensive statistics were gathered about the size, shape, and elemental composition of discharge particles. On the basis of these observations, several hypotheses have been suggested concerning the formation of the particles.

Detectable FDR is mostly particulate in nature. Unjacketed lead bullets produce residue in which greater than 70% of the particles are lead. Coated bullets give the same result, except that a substantial proportion of the lead particles contain copper from the coating material. With jacketed or semi-jacketed bullets the proportion of lead particles in the residue is greatly reduced. It was concluded that most of the lead in the residue comes from the bullet rather than from the primer. This has subsequently been confirmed by experiments involving the use of radioactive tracers.[172]

The stability of the smaller lead particles under the SEM electron beam suggests that they consist of metallic lead rather than lead oxides. Considering that most of the lead particles originate from the bullet, coupled with their stability under the electron beam, leads to the conclusion that the vast majority consist of metallic lead.

Particles due to the discharge of a firearm can be loosely divided into two types: those originating from the bullet, and those originating from the primer. This is a broad general classification only and is not absolute. Particles classified as primer particles because of their elemental content are unlikely to contain metallic lead because they were initially compounds and are unlikely to be reduced to the metal in the oxidizing environment of the primer ignition.

The formation of particles is thought to proceed in the following manner. The hot, high-pressure gases from the burning of the propellant (and primer) act initially on the exposed lead at the base of the bullet. The bullet then passes though the barrel and is subjected to strong frictional heating. This causes some of the bullet and bullet jacket material to be melted and vaporized as well as small fragments to be stripped from the bullet by the rifling. The metal vapors from the bullet mix, to some extent, with the vapors from the inorganic compounds of the primer and are emitted from the firearm through the muzzle and other gaps. They then condense into particles, some of which could be deposited on to the skin and clothing surfaces of the firer.

Discharge particles can be broadly classified into bullet particles and primer particles. This is not unexpected considering that while vapors are miscible, most inorganic compounds in the liquid and solid state cannot dissolve metals, and vice versa. Upon solidification the metals would be expected to separate from the compounds and form separate particles.

Copper is found in both bullet and primer discharge particles if the bullet is coated or jacketed with copper or copper-containing alloy. Discharge residue from ammunition with bare lead bullets shows a very marked decrease in the number of copper-containing particles compared to ammunition with coated or jacketed bullets. This is strong evidence that the copper originates primarily from the bullet coating or jacket rather than from the primer cup assembly or the cartridge case. To explain why copper is found in both bullet and primer discharge particles it is assumed that some proportion of the metallic vapors become oxidized by oxygen or sulfur from the vaporized primer mixture and possibly by atmospheric oxygen outside the gun. Thus the primer (compound) particles contain a contribution from the bullet.

Experiments were conducted involving the addition of tracer compounds to the propellant, followed by an examination of the discharge particles to determine if any of the particles and which type contained the tracer.[173] The tracer was found in primer particles only, which supports the proposition that they should be found dissolved in the oxides and salts originating from the primer and not in the bullet particles. A similar experiment involving the coating of bullets with metals not normally associated with firearms ammunition, followed by examination of the discharge particles, supported the proposition that the bullet material makes a contribution to the primer particles.

It was also observed that the number of discharge residue particles on the firing hand decreased markedly as the bullet velocity increased. One possible explanation for this is that the greatly increased suction in the wake of the faster bullet causes more particles to be sucked out of the muzzle leaving fewer to exit from other vents and to be deposited on the hand. This explanation is supported by the observation that the number of bullet particles deposited is reduced slightly as a function of velocity than is the number of primer particles. Primer particles are on average larger and are produced farther away from the bullet than the bullet particles.

Very little is currently known about the formation and nature of organic discharge residue as the main thrust, until relatively recently, has been directed toward the inorganic content of discharge residue.

Morphology and Size

Apart from various gases four types of particles have been detected in firearm discharge residue:

1. Spherical particles
2. Irregular particles
3. Clusters of particles
4. Fragments of smokeless powder

As a general rule greater than 70% of the particles are spherical. They may be perfect spheres or they may be distorted in some way. Three-dimensional roundness is the basis for this classification. The surface of the particles may be smooth, fuzzy, scaly, or have smaller spheres on their surface. Sometimes they are perforated, capped, broken, or stemmed. The vast majority have diameters of less than 5 µm although they range in size from <0.5 µm to >32 µm (it is not practical to detect particles less than 0.5 µm in diameter).

The irregular particles constitute up to about 30% of the total particle population. The size varies over a wide range from <1 µm to several hundred microns. The larger of the irregular particles sometimes have some small spherical particles attached to them. The irregular particles have the same compositions as the spherical particles.

None of the particles exhibits any features that would suggest a crystal or mineral origin, such as straight or sharp edges, and they are frequently squashed or flattened in appearance.

Clusters consist of from five to several hundred spherical particles attached to each other in a similar fashion to a bunch of grapes. They occur infrequently and appear to be a product of high power or high velocity. Flakes of smokeless powder are few in number and are occasionally seen in promptly collected residue. Unlike the other three types of particles, which are inorganic in nature, the flakes are organic although sometimes spherical particles are embedded in their surface. They range in size from about 50 to 1,000 µm. Clusters and powder flakes are rarely seen in casework as they are usually relatively large in size and would be lost rapidly from skin and clothing surfaces.

Spherical particles are thought to result by rapid condensation from a vapor whereas the irregular particles may be produced by solidification of droplets of molten material that are flung against the inside surfaces of the firearm.

Composition

The identification of a particle as FDR is based on a combination of morphological and elemental composition criteria and also on the association of the particle with other particles present in the sample.

The following three particle compositions have thus far been observed only in FDR and are considered unique to it:

1. Lead, antimony, barium
2. Barium, calcium, silicon, with a trace of sulfur
3. Antimony, barium

The following five particle compositions are consistent with FDR but are not unique to it:

1. Lead, antimony
2. Lead, barium
3. Lead
4. Barium if sulfur is absent or only a trace
5. Antimony (rare)

The lead, antimony and the lead, barium particles, although not unique, have been found in few occupational residues and are therefore considered to be fairly characteristic.

Any particle, unique or consistent, may also contain one or several of the following and only the following elements: silicon, calcium, aluminum, copper, iron, sulfur, phosphorus (rare), zinc (only if copper > zinc), nickel (rare, only with copper, zinc), potassium, chlorine. The presence of some tin is a possibility in obsolete ammunition.

The compositions, shapes, sizes, appearance, and range of particle types should all be considered during interpretation. Particles that are individually consistent with FDR should not be found with otherwise similar particles that are in some way inconsistent with FDR.[170]

The particle classification scheme is the basis of the particle analysis method for FDR detection and identification. A revised particle classification scheme is presented in a later chapter.

Other work has shown that the elemental composition can vary between the inside and the outside of a particle, and using SEM and NAA the average amount of barium and antimony per particle was determined to be in the region of 0.2 to 20 ng for barium and 0.4 to 7 ng for antimony.

Deposition

There is evidence to suggest that the FDR on the firer's hand is blasted on to the hand during the firing process and that residue settling from the atmosphere does not appear to contribute. Copious amounts of FDR issue from the muzzle but play a secondary role in hand deposits. The hand deposits are mainly emitted from openings around the breech and ejection port in self loading firearms and from the flash gap between the cylinder and barrel in revolvers.[174] Such residues can also be deposited on to the face, head hair, and clothing of the firer.

Many factors are thought to influence the amount and nature of FDR deposited on the firer. It should also be borne in mind that there can be a considerable variation in the amount recovered for analysis depending on the sampling method and on how efficiently it is used. Such factors include:

Type of gun—A low success rate for FDR detection has been observed for casework involving .22" caliber revolvers and rifles, and for shotguns which are usually closed breech weapons, some rifles, namely, bolt action, require manual extraction of the spent cartridge case; consequently the design of the firearm can have an influence on the quantity of FDR deposited.

Mechanical condition of the gun—A gun in poor mechanical condition is likely to have larger gaps in the firing mechanism, thereby allowing more FDR to escape.

Cleanliness of the gun—Contact with a "clean" gun is likely to produce less FDR than contact with a "dirty" gun of the same type.

Type of ammunition—Jacketed bullets produce substantially fewer FDR particles than unjacketed bullets. Primer type (size, composition, burning characteristics) can influence the number of primer particles produced. The temperature and pressure achieved by the burning of the propellant determines the power and velocity of the bullet which can influence the number of particles deposited.

Direction and force of air currents (wind)—Tests involving similar firings with the same gun and ammunition batch, one series conducted indoors and the other outdoors, produced substantially less FDR on the firer in the case of the outdoor tests. This was thought to be due to the effect of the wind. Other climatic factors such as rain, humidity, and temperature could also play a part.

Firing location and duration of exposure—Firing from confined spaces, for example, doorways, small rooms, vehicle interiors, will tend to expose the firer to a more residue-laden environment for a longer than normal period of time; consequently, it will take the FDR longer to disperse and the chances of FDR from the muzzle being deposited are increased.

Nature of surface—Skin condition (dry, moist, natural oils, amount of hair) and nature of clothing (smooth or rough) are thought to influence the quantity deposited.

Other factors include the position of the gun relative to the firer; that is, whether the arm is outstretched or closer to firer at time of discharge, whether firing is single or double handed, whether the firer is sitting, standing, kneeling, or lying down will all affect the surface area available for deposition of FDR. Subsequent manipulation of the gun such as unloading, reloading, cleaning, and picking up spent cartridge cases can yield more FDR than the actual shooting.

Quantity and Composition

Bulk elemental analysis methods have been employed by many workers to do quantitative studies of the amount of residue deposited on the firing hand.

Surveys based on indoor firings of a range of handgun/ammunition combination and sampling of the firing hand using a variety of sampling techniques, followed by analysis using differing sample preparation/analytical methods, gave the following widely varying results for residue collected immediately after firing from one to six shots[170]:

Lead	0.45 to 325.0 µg	Average = 7.810 µg
Antimony	0.01 to 10.1 µg	Average = 0.448 µg
Barium	0.02 to 13.7 µg	Average = 0.828 µg

There is very limited information available concerning the total discharge residue particle population from the firing of ammunition. Available data suggest wide variations in total particle population from repeat tests using the same gun, ammunition from the same batch, the same sampling technique, and sampling promptly.

A series of repeat tests involving the firing of a single round, one handed, with clean hands, and sampling by a standardized procedure gave the following results.[170] The mean of six determinations was 5,315 ± 3,622 (68%) for total particles and the mean for the percentage of primer particles was 50.5 ± 14.4

(29%). In another test series the mean of five determinations was 203 ± 81 (40%) for total particles and 65.4 ± 10 (15%) for the percentage of primer particles.

From a single firing of three types of .38" special caliber ammunition the results shown in Table 17.1, for promptly collected residue, were obtained using the particle analysis method.[170]

The vast majority of the total particle population was due to lead only particles originating from the bullet in the case of the RNL (round nosed lead) ammunition whereas the JHP (jacketed hollow point) ammunition produced a much smaller proportion of lead only particles because the base and side of the bullet are enclosed in the jacket, with the only exposed lead at the nose of the bullet.[175] This partly explains the much larger particle population experienced with unjacketed bullets.

Repeated firings with the same gun do not necessarily yield progressively increasing levels of residue nor does the same gun/ammunition combination firing the same number of shots under apparent similar conditions necessarily yield comparable levels of residue.[176]

It is known that FDR consists of gases and a heterogeneous mixture of particles that contain lead, antimony, and barium either individually or in combination. The size of the particles can vary from <1 to >100 μm.

To explain anomalies in quantity and composition of FDR deposited under very similar conditions, it has been proposed that most of the mass of the elements detected is contained in a few large particles and that a large variation occurs in the number and composition of these large particles recovered from firing to firing. It has also been suggested that the skin becomes saturated with residue and that the blast from subsequent shots dislodges residue from previous shots.

It is obvious that the discharge process and the subsequent deposition of FDR on the firer are both subject to many factors, the overall result of which is an unpredictable amount of FDR being deposited. What is deposited, if anything, has a varying composition. The overall process is very random.

Distribution

Distribution may be defined as "the areas where FDR is deposited and the concentration in those areas." It was thought that rifles and handguns when fired in the normal manner would produce different distribution patterns on the firer, with rifles more likely to deposit FDR on the face than handguns. It was also thought that the distribution on the hands could determine whether or not the test subject had discharged or merely handled a gun. On firing a gun more residues are expected to be deposited on the back of the firing hand than on the palm.

Table 17.1 Particle Population, One Shot, Same Caliber

Cartridge	Total Spherical Particles	Pb Only	Ba Only	Pb, Ba	Pb, Sb	Pb, Sb, Ba	Pb, Cu	Nonspherical Particles	Powder Flakes
125 grain JHP Remington	142	35	35	34	13	19	6	13	18
158 grain RNL Remington	2,664	2,086	66	367	39	106	0	28	4
158 grain RNL Federal	4,551	4,162	0	101	95	193	0	0	3

Properties of Firearm Discharge Residue

Although these initial distribution patterns may be correct under ideal test conditions, loss of residue during the course of normal activity and redistribution by transfer from area to area complicate the issue. With the exception of suicides and dead suspects, the initial distribution pattern will almost certainly have altered markedly between the incident and sampling, which is typically several hours after the event. In practice the circumstances of the case dictate the areas to be examined for FDR.

There are many factors that influence the initial distribution pattern and most have been mentioned when discussing the quantity of FDR deposited. This coupled with the loss of residue with time and activity, and the fact that residues are readily transferred from area to area, means that any interpretation based on distribution needs to be approached with caution. Distribution is of limited practical value, but it has been an important factor in a few unusual cases.

Persistence

The persistence of FDR on the hands has been studied by many workers and while there are wide variations in the literature (1 to 24 hours) the majority opinion is in the region of 1 to 3 hours from the event to the sampling of the suspect.[177–179] Beyond 3 hours it is unlikely that residue will be detected on the hands of live suspects.

Persistence on face, head hair, and clothing has not been studied, but it is highly likely that FDR will persist on these areas for longer periods than on the hands.

Persistence data from casework experience are presented later.

Sample Collection

Some form of kit containing sampling materials is normally prepared (commercially, or by police forces, or by forensic laboratories) for the purpose of sampling suspects for FDR.[180–182]

Kit design considerations include (a) avoiding contamination from external sources and cross contamination between sampling areas, (b) lifting efficiency, (c) matrix compatibility with subsequent laboratory procedures, (d) ease of use and preparation, and (e) cost, purity, availability of materials.

Hand sampling methods include:

Cloth	
Cotton wool and dilute acid	Swabbing
Filter paper	
Dilute acid wash	Washing
Paraffin "glove"	Film Lifts
Film forming polymers	
Adhesive tape	Adhesive Lifts
Adhesive stubs	

The analytical technique used strongly influences the type of sampling procedure. Sampling methods fall into two categories: destructive or nondestructive, depending on the effect of the sampling procedure on individual FDR particles, acids tending the break down the particles.

Whatever form of kit is used the lifting efficiency depends to a large extent on the care taken by the sampler.

References

170. G. M. Wolten, R. S. Nesbitt, A. R. Calloway, G. L. Lopel, and P. F. Jones, "Final Report on Particle Analysis for Gunshot Residue Detection," The Aerospace Corporation, El Segundo, CA. Aerospace report no. ATR-77 (7915)-3 (September 1977), 13.
171. G. M. Wolten, and R. S. Nesbitt, "On the Mechanism of Gunshot Residue Particle Formation," *Journal of Forensic Sciences* 25, no. 3 (July 1980): 533.
172. M. A. Purcell, "Radiotracer Studies of Test-Fired Bullets" (master's thesis, University of California, Irvine, 1976).
173. Wolten and Nesbitt, "On the Mechanism of Gunshot Residue Particle Formation," 533.
174. Wolton et al., "Final Report on Particle Analysis for Gunshot Residue Detection," 46.
175. G. M. Wolten, "Cooperative Gunshot Residue Study," *Newsletter* no. 2 (March 31, 1976).
176. R. Cornelis, and J. Timperman, "Gunfiring Detection Method Based on Sb, Ba, lead, and Hg Deposits on Hands. Evaluation for the Credibility of the Test," *Medicine, Science, and the Law* 14, no. 2 (April 1974): 98.
177. J. W. Kilty, "Activity after Shooting and Its Effect on the Retention of Primer Residue," *Journal of Forensic Sciences* 20, no. 2 (1975): 219.

178. R. S. Nesbitt, J. E. Wessel, G. M. Wolten, and P. F. Jones, "Evaluation of a Photoluminescence Technique for the Detection of Gunshot Residue," *Journal of Forensic Sciences* 21, no. 3 (1976): 595.
179. R. Cornelis, and J. Timperman, "Gunfiring Detection Method Based on Sb, Ba, Pb, and Hg Deposits on Hands. Evaluation for the Credibility of the Test, *Medicine, Science, and the Law* 14, no. 2 (April 1974): 98.
180. J. A. Goleb, and C. R. Midkiff, Jr., "Firearms Discharge Residue Sample Collection Techniques," *Journal of Forensic Sciences* 20, no. 4 (1975): 701.
181. M. Tassa, N. Adan, N. Zeldes, and Y. Leist, "A Field Kit for Sampling Gunshot Residue Particles," *Journal of Forensic Sciences* 27, no. 3 (1982): 671.
182. K. K. S. Pillay, W. A. Jester, and H. A. Fox III, "New Method for the Collection and Analysis of Gunshot Residues as Forensic Evidence," *Journal of Forensic Sciences* 19, no. 4 (1974): 768.

Experimental V

18 Objectives, Sampling Procedures, Instrumentation, and Conditions

The purpose of the experimental work was to improve systems for FDR detection and identification, with particular reference to the Northern Ireland situation. This involved looking at all aspects including suspect handling, sampling procedures, laboratory preparation, analysis techniques, interpretation, and presentation of results in a court of law. There is very little information in the literature about suspect handling, contamination avoidance, interpretation of results, and presentation of evidence in court. By taking an overview and a practical approach to all aspects of the system, it is hoped that by introducing improvements, however minor, in each area, these will have a cumulative effect, leading to a substantial overall improvement. This is a novel approach, the vast majority of the literature dealing with scientific methods of sampling, detection, and identification, which while extremely important is not the end product; the end product is the value and credibility of evidence given in court. The main areas for consideration were as follows.

At the start of 1978 the particle analysis method[183] replaced the flameless atomic absorption bulk elemental method[184] as the firearm residue detection method in the NIFSL. Since then the particle analysis method has been substantially improved by the use of a sample concentration/cleanup procedure,[185] the addition of a backscattered electron detector, and the development of an automated residue detection system.[186,187] Despite these improvements the technique remains costly and labor intensive. Certain aspects of the system required further work, in particular, the particle classification scheme: discharge particles from mercury fulminate–primed ammunition and discharge particles from new primer types (Sintox).

The validity of the particle classification scheme was tested by examining items that may produce similar particles, paying particular attention to blank cartridges, the main uses of which that are likely to be encountered in casework are cartridge tools and blank firing replica/imitation firearms.

The particle classification scheme is based on modern primed ammunition and consequently mercury fulminate–primed ammunition is not included. Mercury-containing particles from the discharge of mercury fulminate–primed ammunition are rarely detected in casework. Discharge residue from such ammunition was tested in an effort to provide an explanation for this. Discharge particles from Sintox-primed ammunition was also examined with a view to anticipating future problems the criminal use of this new ammunition may cause for the particle analysis method, range of fire estimations, and the identification of bullet holes.

A major review of 23 years of casework data was undertaken for the following reasons:

- To check the validity of the literature review on the chemistry of ammunition, much of which is based on information gathered over a 23-year period from numerous sources, many of which are of a nonscientific nature, for example, gun magazines, newspaper articles, manufacturers' sales literature.
- To demonstrate the variations in and complexity of the basic item involved in firearms crimes, the round of ammunition, a detailed knowledge of which can aid both the physical and chemical investigation of scenes of crime and the subsequent laboratory examination.
- To record information gained from a terrorist campaign lasting 26 years, some of which will be of interest and benefit to the scientific community and will not be found published elsewhere.
- To clarify the list of accompanying elements in the particle classification scheme and the levels at which they are found.
- To provide an insight into the types of mercury-containing particles detected in casework.

With the emphasis on quality all systems were explored, both internal and external, with a view to ensuring that they could withstand close scrutiny from any source, that the possibility of cross contamination of suspects with explosives and/or firearm discharge residue is minimized, and that contamination risks within the laboratory are identified and minimized or eliminated.

The detection and identification of the organic constituents in FDR has the potential to be used either as a screening technique or, much more likely, as a complementary technique to the particle analysis method. The particle analysis method has proved very satisfactory and has been well tried and tested in casework and court. The objective is to devise an efficient system for organic firearm residue detection that is entirely compatible with the particle analysis method. As a suspect may need to be examined for both firearm and explosive residue the method must also be compatible with organic explosive residue detection techniques.

Summary of aims:

Improve the particle classification scheme.
Explain the scarcity of discharge particles containing mercury.
Clarify the types of discharge particles containing mercury that have been detected in casework.

- Gain information about Sintox ammunition in anticipation of its use in crime.
- Increase knowledge about the chemistry of firearms by reviewing 23 years of casework results and related laboratory tests.
- Improve suspect processing procedures and contamination avoidance measures.
- Devise a method to enable organic FDR detection to be readily incorporated into our existing systems for inorganic FDR and organic explosive residue detection.
- Substantially improve the overall system for firearm and explosive residue detection from the initial arrest of the suspect to the presentation of evidence in court.

Note: For convenience the elements lead, antimony, and barium are referred to as the primary elements in FDR particles. Thus, a single primary element particle would be termed lead only, antimony only, or barium only but it could have, and typically does have, other elements present in the particle from the list of permitted additional accompanying elements.

The terms major, minor, and trace level are defined below and when recording the analysis of a particle all the elements present at a particular level are listed in order of descending peak height. The terms major, minor, and trace are defined in terms of peak height rather than concentration. The strongest peak height should be "on scale" and background levels must be allowed for. The peak heights depend on sample surface irregularities and matrix effects and there is a further complication with overlapping peaks. With this in mind the terms are defined as follows:

Major: Any element whose main peak height is greater than 1/3 of the peak height of the strongest peak in the spectrum.

Minor: Any element whose main peak height is between 1/10 and 1/3 of the peak height of the strongest peak in the spectrum.

Trace: Any element whose main peak height is less than 1/10 of the peak height of the strongest peak in the spectrum.

Instrumentation

Scanning Electron Microscopy

SEM/EDX analyses were performed using a Camscan series 2 scanning electron microscope connected to a Link AN 100000 energy dispersive X-ray analyzer. For automatic detection an automatic residue detection system

(ARDS), developed at this laboratory, was employed the details of which are given in reference 187. For manual operation the following conditions apply:

- 29 mm working distance
- 1-1.25 × 100 µA emission
- × 1,000 magnification
- 25 kV accelerating voltage
- 0° tilt and resolution and absorbed current set according to the microscope on which the examination is carried out (absorbed current 1.84 A)
- Brightness/contrast set on iron particle until just visible on the backscattered image
- Resolution set at spot size 3

Swab and Deldrin samples for SEM/EDX examination were subjected to a concentration/cleanup procedure, the apparatus for which is given in reference 185. All samples were carbon-coated using a Biorad E6430 automatic vacuum controller before examination in the SEM.

Gas Chromatography/Thermal Energy Analyzer

Details of the gas chromatography/thermal energy analyzer (GC/TEA) system are as follows. The instrumentation is a Hewlett Packard 5890 GC with a Thermedics 543 Thermal Energy Analyzer connected to a Hewlett Packard 3393A integrator. The system also incorporates a Hewlett Packard 7673A autoinjector. The TEA was modified according to the alterations detailed in reference 188. The conditions are:

Column:	15M RTX 1 DM Silicon oil, 0.25 mm internal diameter
Initial temperature:	100°C for 3 minutes
First ramp:	35°C to 165°C, held at 165°C for 3 minutes
Second ramp:	30°C to 195°C to final temperature 195°C for 3 minutes
Carrier gas:	Helium
Flow:	2.5 ml/min
Injection temperature:	198°C
Interface temperature:	250°C–TEA
Pyrolyzer temperature:	800°C–TEA
Injection volume:	5 µl
Specificity:	All N-nitroso compounds plus organic and inorganic nitrites, some nitrates, certain alkyl C–nitroso compounds and certain C-nitro compounds

Sensitivity: Typically 100 pg in 5 µl injection
Linearity: 0.04 to 0.6 ng/µl
Precision: 1 standard deviation ± 0.01 min; coefficient of variation typically 0.15%

High-Performance Liquid Chromatography/ Pendant Mercury Drop Electrode

The HPLC/PMDE instrumentation, materials, conditions and performance data are detailed in reference 189.

Gas Chromatography/Mass Spectrometry

The GC/MS instrumentation, materials, and conditions are detailed in reference 188, and performance data for a typical analysis of 1 ng/µl standards are:

Standard	Mean Retention Time (min)	Standard Deviation	Coefficient of Variation (%)
Camphor	2.580	0.008500	0.3290
Nitroglycerine (NG)	4.112	0.019050	0.4630
1,3-Dinitrobenzene (1,3-DNB)	4.905	0.064280	1.3100
2,4-Dinitrotoluene (2,4-DNT)	5.709	0.072230	1.2700
Diphenylamine (DPA)	6.581	0.008869	0.1350
Methylcentralite (MC)	8.213	0.015760	0.1920
Ethylcentralite (EC)	8.672	0.008180	0.0943

Flameless Atomic Absorption Spectrophotometry

Lead, antimony, barium, mercury, and copper determinations were done using the instrumentation, materials, and methods detailed in reference 184. All other elements were analyzed using the standard conditions as detailed in the instrument manual.

Sampling

Cartridge Cases

The interior of spent cartridge cases were sampled by dry swabbing with ~0.25 g of acrilan fiber for FAAS analysis (dry swabbing) or by double-sided adhesive tape (Scotch pressure sensitive tape) on the end of a suitable diameter

Perspex rod (tape lift) for SEM/EDX analysis; the tape was then transferred to a SEM sample stub for examination.

FDR/Explosive Residue

Unless stated otherwise, hands were sampled by tape lift (adhesive stub) for inorganic FDR using a 13-mm-diameter aluminum SEM sample stub with double-sided adhesive tape attached (Scotch pressure sensitive tape). Sampling of the hands and face in the outdoor firing tests in Chapter 26 employed acrilan fiber damped with isopropanol (IPA) as per the suspect sampling kit detailed in reference 190.

Unless stated otherwise, clothing was sampled using the apparatus detailed in Figure 26.1 and Figure 26.2 (suction sampled).

References

183. G. M. Wolten, R. S. Nesbitt, A. R. Calloway, G. L. Lopel, and P. F. Jones, "Final Report on Particle Analysis for Gunshot Residue Detection," The Aerospace Corporation, El Segundo, CA. Aerospace report no. ATR-77 (7915)-3 (September 1977).
184. J. S. Wallace, "Firearms Discharge Residue Detection Using Flameless Atomic Absorption Spectrophotometry," *AFTE Journal* 19, no. 3 (July 1987).
185. J. S. Wallace, and R. H. Keeley, "A Method for Preparing Firearms Residue Samples for Scanning Electron Microscopy," *Scanning Electron Microscopy* 2 (1979).
186. T. G. Kee, C. Beck, K. P. Doolan, and J. S. Wallace, "Computer Controlled SEM Micro Analysis and Particle Detection in the Northern Ireland Forensic Science Laboratory—A Preliminary Report," Home Office Internal Publication. Technical note No. Y 85 506.
187. T. G. Kee, and C. Beck, "Casework Assessment of an Automated Scanning Electron Microscope/Microanalysis System for the Detection of Firearms Discharge Particles," *Journal of the Forensic Science Society* 27 (1987): 321.
188. J. M. F. Douse, "Trace Analysis of Explosives at the Low Nanogram Level in Handswab Extracts Using Columns of Amberlite XAD-7 Porous Polymer Beads and Silica Capillary Column Gas Chromatography with Thermal Energy Analysis and Electron Capture Detection," *Journal of Chromatography* 328 (1985): 155.
189. S. J. Speers, K. Doolan, J. McQuillan, and J. S. Wallace, "Evaluation of Improved Methods for the Recovery and Detection of Organic and Inorganic Cartridge Discharge Residues," *Journal of Chromatography* 674 (1994).
190. J. S. Wallace, and W. J. McKeown, "Sampling Procedures for Firearms and/or Explosives Residues," *Journal of the Forensic Science Society* 30 (1993): 107.

Particle Classification Scheme 19

Blank Cartridges

Doubts initially arose about the particle classification scheme when the NIFSL experienced a case involving the examination of a suspect's upper outer clothing for FDR, in which the only discharge particle types detected were lead only and barium, calcium, silicon (the barium, calcium, silicon type was considered to be unique to the discharge of a firearm at the time). The size and appearance of the particles were consistent with FDR residue; however, none of the other particle types were detected and test firing of similar ammunition to that used in the incident (same caliber and head stamp) produced the complete range of particle types. In an effort to explain this anomaly it was decided to investigate the suspect's occupation as a possible source of the particles. Inquiries revealed that the suspect was employed as a general laborer on a building site and that cartridge-operated industrial tools (stud guns) were used on that site. Consequently it was necessary to investigate discharge residue particles from all types of cartridge-operated industrial tools used in Northern Ireland, in order to establish if this was the source of the particles detected on the suspect. The results of the study[191] neither proved nor disproved that the use of a cartridge tool accounted for the particles on the suspect, but it did show that the particle classification system needed revision.

Two other similar cases were subsequently experienced and as a consequence it was decided to act on the side of safety by reclassifying barium, calcium, silicon particles as indicative, rather than unique, to the discharge of a firearm.

To clarify the situation and to test the validity of the particle classification scheme it was necessary to consider other possible sources of particles that may have a similar elemental composition and morphology to FDR particles. As the formation of FDR particles involves high temperatures and the particles have the appearance of condensing from a vapor or melt, it was decided to study blank cartridges other than those used in cartridge tools. A very limited range of toy caps, matches, signal flares, and fireworks were also examined as these, when used, involve high temperatures and may contain one or more of the elements lead, antimony, or barium, elements associated with firearms ammunition.

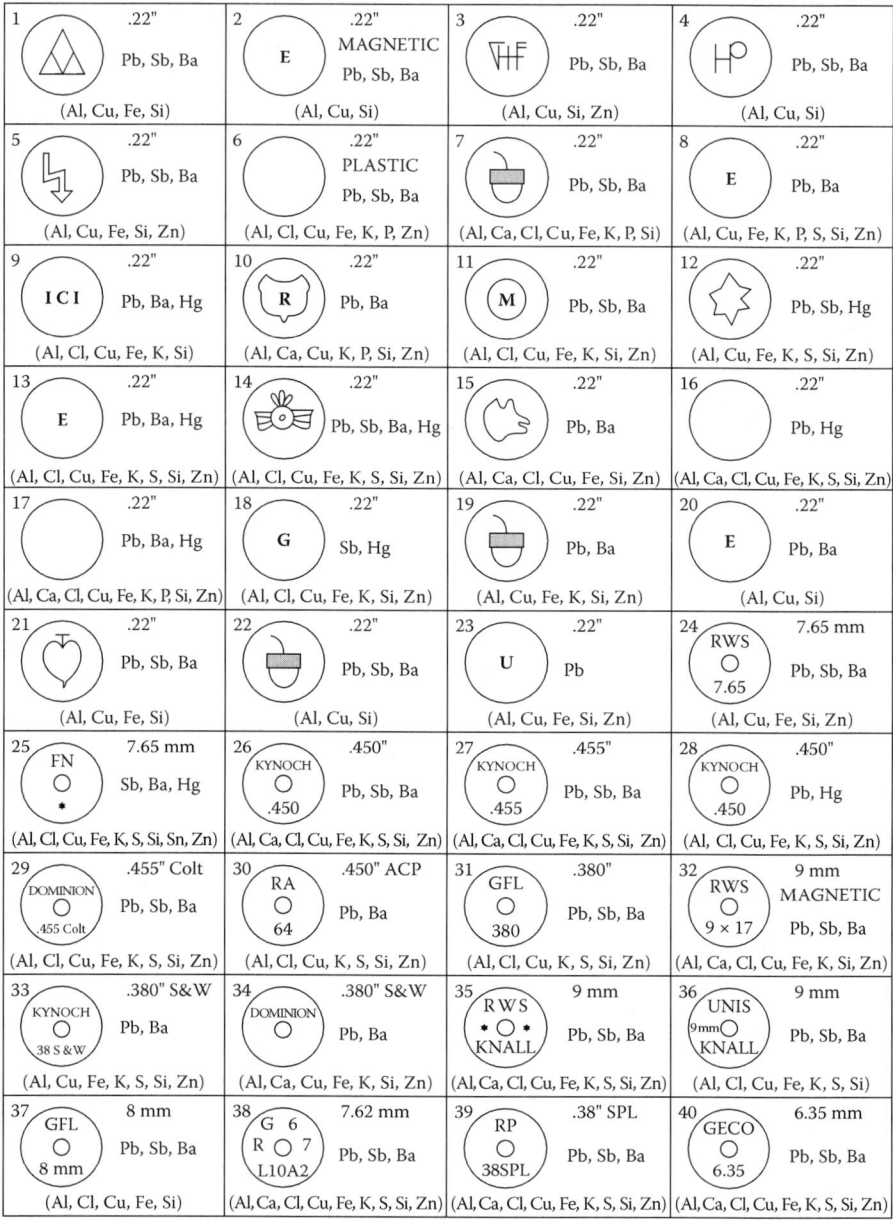

Figure 19.1 Residue in spent cartridge case (blanks).

A range of blank cartridges were examined to determine the elemental content of the primer. Particular attention was given to blank cartridges designed for use in starting pistols and replica firearms as these sources are the most likely to be encountered in casework. Results are given in Figure 19.1.

Particle Classification Scheme

#	Label	Caliber	Elements	Trace
41	GECO 7 O 7 6.35	6.35 mm	Pb, Sb, Ba	(Al, Cu, Fe, K, Si)
42	7.62 × 57 O DAG 66-64	7.62 mm PLASTIC	Pb, Sb, Ba	(Al, Ca, Cl, Cu, Fe, Si)
43	11 T O T 62	7.5 mm	Pb, Sb, Ba	(Al, Ca, Cl, Cu, Fe, K, Si, Zn)
44	9 × 19 O GECO 6-6	9 mmP	Pb, Sb, Ba	(Al, Cu, Cl, Fe, Si, Zn)
45	K O 9 mm	9 mmP	Pb, Sb, Ba	(Al, Ca, Cl, Cu, Fe, K, Si, Zn)
46	⊕ DNG-68-6	9 mmP	Pb, Sb, Ba	(Al, Cu, Fe, K, Si, Zn)
47	AMA O 57	9 mmP	Pb, Sb, Ba	(Al, Ca, Cu, Fe, Si, Zn)
48	S.F.M ☆O☆	9 mm	Sb, Ba, Hg	(Al, Cu, Fe, K, S, Si, Zn)
49	RG O 54 L10Z	.303"	Pb, Sb	(Al, Cl, Cu, Fe, K, Si, Zn)
50	74 O RG Z67	.303"	Pb, Ba	(Al, Ca, Cl, Cu, Fe, K, Si)
51	B↑E O 42 H1Z	.303"	Pb, Sb, Hg	(Al, Ca, Cl, Cu, Fe, K, S, Si, Zn)
52	SL O 53	.30-06	Pb, Sb, Ba	(Al, Cl, Cu, Fe, K, Si)
53	MCM ( ) 320	.320"	Sb, Hg	(Al, Cl, Cu, K, S, Si, Zn)
54	RWS O .320	.320"	Pb, Sb, Ba, Hg	(Al, Cl, Cu, Fe, K, Si, Zn)
55	VFM&C O .32 LIÈGE	.320"	Pb	(Al, Cl, Cu, Fe, K, S, Si, Zn)
56	G.F.L O .320	.320"	Pb, Sb, Ba	(Al, Cl, Cu, Fe, K, Si)
57	M I N O O O D O N 32 S&W	.320" S&W	Pb, Ba	(Al, Ca, Cl, Cu, Fe, K, Si, Zn)
58	N O C Y O H K O H .32 S&W	.320" S&W	Pb, Ba, Hg	(Al, Ag, Ca, Cl, Cu, Fe, K, S, Si, Zn)
59	N O C Y O H K O H .32 S&W	.320" S&W	Pb, Sb, Ba	(Al, Cl, Cu, Fe, K, Si, Zn)
60	O	.320"	Sb, Hg	(Al, Cl, Cu, Fe, K, Si, Zn)
61	TW O 73	.223"	Pb, Sb, Ba	(Al, Cl, Cu, Fe, K, Si, Zn)
62	O	.30-06 PLASTIC	Pb, Sb, Ba	(Al, Ca, Cl, Cu, Fe, S, Si, Zn)
63	LC O 53	.30-06	Pb, Sb, Ba	(Al, Ca, Cl, K, Cu, Fe, S, Si, Zn)
64	K66 O .455	.455"	Hg	(Al, Ag, Cl, Cu, K, Si, Zn)
65	M I N O O O D O N .455 COLT	.455" COLT	Pb, Ba	(Al, Ca, Cl, Cu, Fe, K, Si, Zn)
66	K 66 O L5A3	.450" ACP	Pb, Sb, Ba	(Al, Ca, Cl, Cu, Fe, K, S, Si, Zn)
67	DAG O 9mm PARA	9 mmP	Pb, Sb, Ba	(Al, Cl, Cu, Fe, K, Si, Zn)
68	WW 38 O SPECIAL	.38 SPL	Pb, Sb, Ba	(Al, Cl, Cu, Fe, K, Si, Zn)
69	RWS O 9 × 17	.380"	Pb, Ba	(Al, Ca, Cl, Cu, Fe, K, Si, Zn)
70	R.P O 38 SPL	.38" SPL	Pb, Sb, Ba	(Al, Cl, Cu, Fe, K, Si, Zn)
71	RWS ♡ 38 S&W	.38" S&W	Pb, Ba	(Al, Ca, Cl, Cu, Fe, K, Si, Zn)
72	GECO O 8 mm	8 mm	Pb, Sb, Ba	(Al, Cl, Cu, Fe, K, Si, Zn)
73	E O	.22" LR	Pb, Ba	(Al, Cl, Cu, Fe, K, S, Si, Zn)
74	M I N O O O D O N 32 S&W	.32" S&W	Pb, Ba	(Al, Ca, Cl, Cu, Fe, K, Si, Zn)
75	RWS O 320	.320"	Pb, Ba	(Al, Ca, Cl, Cu, Fe, K, Si, Zn)
76	RG O 75 L13A1	7.62 mm	Pb, Ba	(Al, Ca, Cl, Cu, Fe, K, Si, Zn)
77	TW O 70	.223"	Pb, Sb, Ba	(Al, Cl, Cu, Fe, K, S, Si, Zn)
78	E O	.22" LR	Pb, Ba	(Al, Cl, Cu, Fe, Si, Zn)
79	RG O 38SPL	.38" SPL	Pb, Sb, Hg	(Al, Cl, Cu, Fe, K, Si, Zn)

Figure 19.1 (Continued.)

Note: Comparison of Figure 19.1 with Table 19.1 and Table 19.2 reveals that blank cartridges are similar in composition to live ammunition and may be expected to produce discharge particles with similar composition to those originating from firearm discharge.

Table 19.1 Starting Pistol Discharge Residue

Blank No.	Pb, Sb, Ba	Pb, Sb	Pb, Ba	Sb with S	Ba Only	Pb Only
1	103	ND	2	ND	2	ND
2	96	14	7	ND	ND	ND
3	116	4	ND	5	ND	ND
4	126	8	ND	ND	ND	6
5	80	7	8	10	ND	ND
6	133	ND	ND	ND	ND	ND
7	99	ND	ND	3	ND	ND

ND = none detected.

Table 19.2 Particle Types from Promptly Collected FDR

Particle Type	Unjacketed Bullet .38 Special Caliber		Jacketed Bullet .38 Special + P Caliber	
	Number	Approximate %	Number	Approximate %
Pb, Sb, Ba	39	17.0	29	27.0
Sb, Ba	None	—	None	—
Ba, Ca, Si	3	1.5	None	—
Pb, Sb	42	18.0	16	15
Pb, Ba	6	3.0	44	40.0
Pb only	138	59.0	14	13.0
Sb only	6	3.0	1	0.5
Ba only	None	—	4	4.0
Ratio indicative/unique		5:1		4:1

Discharge residue particles from starter pistol blanks were examined for comparison with discharge residue particles originating from firearms ammunition. Table 19.1 gives the starting pistol discharge residue particles classified according to their consistency with FDR particles.

The particles detected were all in the size range 1 to 19 μm. Both spherical and irregular particles were found and all had the appearance of having partially or wholly condensed from a vapor or melt, and all exhibited some degree of curvature. They did not exhibit any degree of crystallinity and their surface details were smooth, irregular, cratered, or nodular. The physical characteristics of the particles were indistinguishable from FDR particles.

Individually, the starting pistol discharge residue particles could not be distinguished from FDR particles, by physical appearance, by size range, or by elemental composition, which includes the additional accompanying elements. It is a reasonable assumption that discharge residue particles from any blank cartridge could be confused with FDR. Mercury-containing particles could also be produced from the use of blanks incorporating mercury fulminate (see Figure 19.1 for mercury-containing blanks).

When the particles were considered as a group three distinct differences between firearm and starting pistol discharge residue particles were noted.

1. The ratio of indicative to unique particles is markedly different from that of firearm discharge. From firearm residue casework statistics, based on cases with at least one particle in the unique category, the ratio of indicative to unique particles is approximately 35:1. For starting pistol discharge residue particles the overall ratio is in the region of 1:10.

This abundance of unique discharge particles suggests a much more homogeneous mixture of discharge residue, which is not surprising considering that the blanks were rimfire primed; the chemicals are contained within a relatively small volume compared to firearm ammunition, that is, a more intimate mixture contained in a smaller cartridge case. In addition there is no bullet involved to complicate the issue by producing a large number of bullet particles, which would make a substantial contribution to the total number of particles in the indicative category.

It is interesting to note that blank number 6, Table 19.1 produced lead, antimony, barium unique particles only, and also that it had a plastic cartridge case. Considering that eight rounds were fired and that a primer composition consists of a mixture of chemical compounds, this result tends to suggest that the mixture was initially homogeneous and that the discharge gases/vapors were intimately mixed prior to condensation of the particles. This trend is noticeable throughout Table 19.1, particularly for blanks numbered 1, 6, and 7.

2. Unlike FDR, the discharge residue from the blank cartridges contained very few lead-only particles, which is not unexpected as there is no lead bullet involved and any lead-only particles detected must originate from the primer. The few lead-only particles detected all originated from the discharge of blank cartridge number 4. A tentative explanation for this could be the relative proportion/total quantity/burning rate/granulation size of the lead compound or compounds or the degree of uniformity of the priming mixture.

3. Unlike firearm discharge residue, each of the blank cartridges produced a limited range of discharge particle types.

There are several things to be considered when comparing the ratio of indicative to unique particles in firearms casework and starting pistol discharge residue. Like with like is not being compared in that samples of starter pistol discharge residue were taken immediately after firing whereas in casework the vast majority of the suspects were apprehended between 1 and 4 hours after the shooting incident. In laboratory tests one has a high degree of confidence in the origin of the particles in the indicative category, whereas in casework one cannot be sure of the origin of some of the particles,

Table 19.3 Elemental Level Per Particle Type

Particle type	Element	~% Major	~% Minor	~% Trace
Pb, Sb, Ba	Pb	61	39	—
	Sb	39	31	30
	Ba	64	31	5
Sb, Ba	Sb	—	12	88
	Ba	100	—	—
Ba, Ca, Si	Ba	93	7	—
	Ca	28	55	17
	Si	34	52	14
Pb, Ba	Pb	95	5	—
	Ba	38	57	5
Pb, Sb	Pb	66	34	—
	Sb	55	40	5
Sb only	Sb	92	8	—
Ba only	Ba	100	—	—
Pb only	Pb	95	5	—

particularly the single primary element ones. In order to compare like with like a further experiment was conducted involving promptly collected FDR. Results are given in Table 19.2.

Note: Examination of the copper and zinc relationship revealed that for the unjacketed bullet, 94% of the particles contained copper only and 6% contained both copper and zinc, with copper > zinc. For the jacketed bullet, 90% of the particles contained copper only and 10% contained both copper and zinc, with copper > zinc in 9.5% and copper = zinc in 0.5%.

As can be seen from Table 19.3, the proportion of indicative particles exceeds the proportion of unique particles, even for promptly collected FDR. The higher proportion of indicative particles detected in casework is almost certainly due to particles from nonfirearm sources, particularly single primary element ones, meeting the criteria of the classification scheme.

It is interesting to note that the firing of ammunition with an unjacketed bullet produced more lead-only particles than similar ammunition with a jacketed bullet, which is consistent with the findings of the Aerospace Corporation work.[192] A surprising result was the number of particles containing copper from the firing of the unjacketed bullets. This is inconsistent with its findings and is difficult to explain, as the only obvious source of copper is the cartridge case/primer cup. It concluded that these sources did not appear to make a significant contribution to the elemental composition of the discharge particles. Little significance can be attached to this finding as it is based on a particular gun/ammunition combination and very limited experimental data.

Starting pistols/blank firing imitation firearms normally have a hardened steel blockage in the barrel to prevent them from being converted to fire bulleted ammunition. Firearms and firearms ammunition are designed so that the maximum pressure is reached when the bullet has traveled a considerable distance up the barrel (like an expanding chamber). Thus the nature of the discharge process differs between firearms and blank firers and this could account for the homogeneous character of the discharge gases and vapors from blank cartridges. In a firearm the vast majority of the discharge residue emerges from the muzzle whereas blank firers have a small vent, usually at the top, to emit the discharge residue. Because of the smaller fixed volume available to the discharge residue gases and vapors and the venting mechanism in blank firers, it is likely that more uniform temperatures and pressures are attained and better mixing occurs, leading to an abundance of lead, antimony, barium particles and a limited range of particle types.

Whatever the reason, there is no doubt that the discharge of blank cartridges produces a much higher ratio of unique to indicative particles than the discharge of firearm ammunition. As a consequence of the work on blank cartridges, discharge residue particles that were previously referred to as FDR are now referred to as cartridge discharge residue (CDR).

Toy Caps

Six different brand names of caps designed for use in toy guns were examined, two of which were the paper roll type; the others were the individual plastic cup type that is placed on the "anvil" of the toy gun.

Analysis of discharge particles revealed that both spherical and irregular particles were present, with approximately 1 in 12 spherical. The particle size range was from 3 to 160 μm. The elements detected were aluminum, calcium, chlorine, copper, iron, potassium, magnesium, phosphorus, lead, sulfur, antimony, silicon, titanium, and zinc, with calcium, chlorine, potassium, phosphorus, lead, and silicon the major elements. Antimony and lead did not occur together and none of the samples examined would be confused with FDR particles as their elemental profile differed. A small proportion of the particles containing either lead or antimony met the criteria for "single" element FDR particles.

At the time the tests were conducted, children in Northern Ireland played with "devil bombers," which consisted of a solid mixture rolled up in a piece of waxed paper. When thrown with force against a hard object they exploded creating a loud bang. Visual examination of the contents revealed a mixture of woodlike material (cellulose) and sandlike material (silicate). Elemental analysis of the mixture showed silver, silicon at major level, aluminum at minor level, and potassium, chlorine at trace level. Its exact composition was not determined but it would appear that cellulose was a fuel, silicate was a frictionator, a silver compound (azide or fulminate?) was the explosive, and potassium chlorate was the oxidizer.

Matches

Analysis of particles originating from the use of matches revealed that only a very small number of spherical particles were present; the majority of particles was very irregular. The elements detected were aluminum, calcium, chlorine, chromium, iron, potassium, magnesium, manganese, phosphorus, sulfur, antimony, silicon, and zinc, with potassium, chlorine, phosphorus, sulfur, and silicon the major components. Antimony was detected in only 2 of the 17 types of matches examined. None of the samples examined would be confused with FDR particles as both their morphology and elemental content differed.

Flares

Flares have several uses including signaling and illumination and there are several means of launching, including handheld, rocket, and specifically designed pistols (for example, Verey pistol). The use of flares in Northern Ireland is very limited, with the security forces using them occasionally. They have in rare instances been used illegally. Analysis of two handheld types showed that the vast majority of the discharge particles were irregular with several large flakes present. Elemental analysis revealed the presence of calcium, copper, iron, magnesium, sodium, titanium, zinc in one of the flares, with magnesium, sodium at major level, and aluminum, barium, chlorine, iron, potassium in the other, with aluminum, potassium, chlorine at major level. Their morphology and composition was such that they would not be confused with FDR particles.

The flares examined were the only ones used by the security forces at the time. A brief review of the literature[193-195] on pyrotechnics/flares indicates that lead and antimony compounds are infrequently used and when used do not occur together. Barium compounds are frequently used, particularly in signal flares. From the literature it is apparent that residues from flares could not be confused with FDR as the elements lead, antimony, and barium would be accompanied by other elements that would clearly indicate a non-FDR source.

Fireworks

During the troubles in Northern Ireland only indoor-type fireworks could be purchased without a special license. Analysis of particles originating from the use of indoor-type fireworks showed only a few spherical particles; the majority was large irregularly shaped flakes. The elements aluminum, barium, chlorine, chromium, iron, potassium, sulfur, and antimony were detected, all of which were at a major level.

Analysis of particles originating from the use of outdoor fireworks revealed that the majority of the particles was irregular, many were crystalline, and many large flakes were present. A small proportion of the particles were spherical and physically resembled FDR particles. Elemental analysis showed the presence of aluminum, arsenic, barium, calcium, chlorine, copper, iron, potassium, magnesium, sodium, lead, sulfur, antimony, silicon, strontium, titanium, zinc, and zirconium. None of the particles detected would be confused with FDR particles as the primary FDR elements were always accompanied by elements that were clearly of non-FDR source.

In conclusion, lead, antimony, and barium may be encountered in pyrotechnics, in both fireworks and flares. Lead and antimony were present in toy caps but were not found occurring together. Antimony-only was detected in matches. None of these sources should be confused with FDR particles as their morphology and/or elemental content differs. (The text on toy caps, matches, flares, and fireworks represents the conclusions of the work conducted, as the details and results were lost in the terrorist explosion at the NIFSL in September 1992.)

Accompanying Elements

From casework statistics the unique particles (those containing the combination lead, antimony and barium, and those containing antimony and barium) occur in the ratio 7:3, respectively. Approximate percentages for indicative particles are lead-only 55%; lead, antimony 20%; lead, barium 8%; antimony-only 7%; barium, calcium, silicon 5%; barium-only 5%. Table 19.3 gives an indication of the levels of the primary elements in each particle type. Table 19.4 gives an indication of the levels of accompanying elements in each particle type and is the basis for note b in Table 19.5, Particle Classification Scheme.

The work serves to illustrate the heterogeneous nature of firearm discharge residue particles and to clarify the types of particles detected.

Table 19.4 Percentage Occurrence of Certain Accompanying Elements in Unique and Indicative Particles

Element	Pb, Sb, Ba				Sb, Ba				Ba, Ca, Si				Pb, Ba				Pb, Sb				Sb Only				Ba Only				Pb Only			
	Total %	Minor	Major	Trace	Total %	Minor	Major	Trace	Total %	Minor	Major	Trace	Total %	Minor	Major	Trace	Total %	Minor	Major	Trace	Total %	Minor	Major	Trace	Total %	Minor	Major	Trace	Total %	Minor	Major	Trace
Al	47.0	19.5	19.5	8.0			Zero		69.0	17.0	41.5	10.5	35.0	Zero	30.5	4.5	53.0	1.0	24.5	27.5	88.5	Zero	21.0	67.5	87.5	Zero	55.0	32.5	63.0	Zero	26.0	37.0
Ca	86.0	33.0	53.0	17.0	100.0	87.0	13.0	Zero	100.0				97.0	73.0	20.0	4.0	15.5	1.0	10.0	4.5	16.0	6.5	3.0	6.5	91.0	3.0	25.0	63.0	69.0	9.5	37.0	22.5
Cl	43.5	Zero	26.5	17.0	93.0	Zero	11.5	81.5	69.0	Zero	20.5	48.5	36.0	0.5	28.0	7.5	37.5	1.0	6.5	30.0	84.0	Zero	26.0	58.0	92.5	Zero	12.5	80.0	34.0	Zero	11.5	22.5
Cr	Less than 0.5 (all trace)				Zero				Zero				3.0	Zero	2.5	0.5	1.0 (all trace)				Zero				Less than 0.5 (all trace)				1.5 (all trace)			
Cu	96.0	Zero	52.5	43.5	99.0	Zero	50.0	49.0	90.0	Zero	7.0	83.0	99.0	Zero	30.0	69.0	100.0	Zero	22.5	77.5	89.0	Zero	6.5	82.5	47.5	Zero	2.5	45.0	76.5	3.5	18.5	54.5
Fe	94.5	1.5	57.5	35.5	100.0	Zero	93.0	7.0	79.5	3.5	41.5	34.5	99.5	Zero	91.5	8.0	73.0	Zero	36.5	36.5	92.0	Zero	14.5	77.5	37.5	Zero	20.0	17.5	64.5	Zero	22.0	42.5

Particle Classification Scheme

	1	2	3	4	5	6	7	8
K	58.0 / Zero / 52.5 / 5.5	Zero	65.5 / Zero / 27.5 / 38.0	47.5 / Zero / 35.5 / 38.0	36.0 / 1.0 / 26.0 / 9.0	24.0 / Zero / 9.5 / 14.5	89.0 / Zero / 4.0 / 85.0	50.0 / Zero / 14.0 / 36.0
Mg	1.0 (all trace)	Zero	14.0 (all trace)	28.0 / Zero / 19.5 / 8.5	20.0 (all trace)	9.5 (all trace)	Zero	12.0 (all trace)
Na	4.0 (all trace)	Zero	Zero	1.0 (all trace)	Zero	0.5 (all trace)	Zero	1.0 (all trace)
P	3.0 / Zero / 1.5 / 1.5	Zero	14.0 / Zero / 3.5 / 10.5	6.0 / Zero / 3.0 / 3.0	7.5 (all trace)	15.0 / Zero / 2.0 / 13.0	Zero	23.0 / Zero / 10.0 / 13.0
S	67.0 / 31.5 / 35.5 / Zero	70.0 (all minor)	59.0 (all trace)	85.5 / 81.0 / 4.5 / Zero	37.5 / 10.5 / 25.5 / 1.5	81.0 / 14.5 / 40.5 / 26.0	85.0 (all trace)	27.0 / 24.0 / 3.0 / Zero
Si	97.0 / 20.5 / 64.5 / 12.0	100.0 / 81.5 / 18.5 / Zero	100.0	99.0 / 76.0 / 23.0 / Zero	95.5 / 12.0 / 66.5 / 17.0	100.0 / 11.5 / 30.5 / 58.0	25.5 / 5.0 / 3.0 / 17.5	83.5 / 12.0 / 45.0 / 26.5
Ti	5.0 (all minor)	Zero	1.0 (all trace)	4.5 (all trace)	7.0 / Zero / 2.0 / 5.0	16.0 / Zero / 1.0 / 15.0	3.5 / Zero / 0.5 / 3.0	10.0 / Zero / 3.0 / 7.0
Zn	7.5 / Zero / 5.0 / 2.5	18.0 (all minor)	14.0 (all trace)	27.5 / Zero / 21.0 / 6.5	7.5 (all trace)	7.0 (all trace)	7.5 / Zero / 1.0 / 6.5	7.0 / Zero / 4.5 / 2.5

Table 19.5 Particle Classification Scheme

Unique[b,c]	Indicative[b,c]
Pb, Sb, Ba	Ba, Ca, Si[a]
Sb, Ba	Pb, Ba
	Pb, Sb
	Sb only (with S)
	Ba only[a]
	Sb only (without S)
	Pb only

[a] S absent or acceptable at trace level only when Ba is present at a major level.

[b] Any of the particles listed may also include some of the following: Al, Ca, Si, S (unless specifically excluded) at *Major, minor, or trace:* Cl, Cu, Fe, K, P, Zn—only if Cu also present and Cu > Zn at *Minor or trace:* Co, Cr, Mg, Mn, Na, Ni, Ti (typically none present, occasionally one, rarely two) at *Trace only.*

The presence of Sn suggests mercury fulminate–primed ammunition. (Sn is present in some propellants; it has been used to harden bullet Pb and it is present in some bullet jackets.)

Particle Classification Scheme

The original particle classification scheme[192] has been revised based on casework experience, research work on blank cartridges, and so forth, and a detailed analysis of 14 years of casework results. The particle classification scheme used in Northern Ireland since 1984 is given in Table 19.5. The indicative particles are in tentative order of decreasing significance.

The classification scheme is based on discharge residue particles from modern primed brass-cased ball ammunition. It is only applied rigidly when no other information is available. When a gun, ammunition, spent cartridge case, or bullet is recovered, it can be examined to determine elemental composition and likely discharge residue particle composition.

The classification scheme has to be flexible in order to encompass the wide range of different primer/cartridge case/propellant/bullet combinations. For example, zinc-coated steel-cased ammunition gives iron and zinc at major levels in the discharge particles; firearms with rusted barrel interiors or the use of steel jacketed bullets can produce discharge particles with iron at major level; primers containing lead hypophosphite can give discharge particles with phosphorus at major level; ammunition with black powder can produce discharge particles with potassium and sulfur at major level. Because of these and other variables the classification scheme has to be flexible.

It must be stressed that the classification scheme is intended as a general guide and is only applied rigidly when there is nothing recovered for comparison purposes.

FDR particles have been noted in a wide range of shape, size, and appearance. They all have the appearance of having condensed from a vapor or melt, namely, a three-dimensional roundness. Ragged or straight edges or corners suggest a mineral origin. The shape and appearance is particularly important in the indicative category to aid the differentiation from occupational/environmental particles. (Particles originating from the bullet/bullet jacket are sometimes encountered. These are usually identifiable and are not included in the particle classification scheme.)

References

191. J. S. Wallace, and J. McQuillan, "Discharge Residues from Cartridge-Operated Industrial Tools," *Journal of the Forensic Science Society* 24 (1984): 495.
192. G. M. Wolten, R. S. Nesbitt, A. R. Calloway, G. L. Lopel, and P. F. Jones, "Final Report on Particle Analysis for Gunshot Residue Detection," The Aerospace Corporation, El Segundo, CA. Aerospace report no. ATR-77 (7915)-3 (September 1977).
193. R. Harris, "Pyrotechnic Compositions," *Chemistry in Britain* (March 1977): 113.
194. *Kirk-Othmer Encyclopedia of Chemical Technology,* 2nd ed., vol. 13 (New York: Wiley-Interscience), 824.
195. Private communications, 1983.

Casework-Related Tests 20

Particles from Handling Ammunition

Particles on the outside surface of newly acquired, unopened ammunition were examined in order to determine if the ammunition was contaminated with discharge residue in the factory. Munitions and firearms manufacturers do test-fire their products. Results are given in Table 20.1.

The majority of the lead-only, antimony-only, and lead, antimony particles that were spherical would be classified as indicative of FDR. However, they were accompanied by particles whose morphology was inconsistent, and only a limited range of particle types were present. No unique FDR particles were detected.

A further test was conducted to determine whether or not ammunition that had been previously loaded in a firearm would have FDR on its surface. Results are given in Table 20.2.

Table 20.2 shows that the complete range of FDR types can be deposited from handling ammunition that has been in a firearm. It is reasonable to assume that the same applies to handling magazines, or ammunition that has been in a magazine. Particles similar to those detected in Table 20.1 were also present. The presence of FDR on a suspect's hands could arise from handling ammunition that had been chambered in a firearm or from handling spent cartridge cases, a gun, or a magazine. Consequently, the presence of FDR on the hands does not prove that the suspect fired a gun, but does infer recent involvement with firearms or related items.

Bullet Weight Loss on Firing

A test was conducted to determine the weight loss of some bullets after discharge. Results are given in Table 20.3. From the limited experimental data it would appear that, as expected, the full metal jacketed bullets lose less than the soft unjacketed bullets. The FMJ bullet with its base enclosed lost less than its equivalent with its base exposed. This is also predictable as the exposed base is subject to erosion during discharge. The .38 SPL + P unjacketed bullet

Table 20.1 Particles on New Ammunition

Ammunition	Pb Only	Sb Only	Pb, Sb	Brass	Observations
GECO 9 mm Luger. Brass-jacketed bullet	Large number	None	None	Large number	All the brass particles were irregular shaped whereas the Pb-only particles were a mixture of irregular and spherical. The Pb-only particles contained some or all of Al, Ca, Cl, Cu, S, Si, Ti, Zn at minor or trace level. A few Fe particles were present. Unusual particles detected were Bi, P, Al major, Si minor, Ca, Fe trace; Cr, Fe major, Si, Zn trace; Zn major, Fe, Cu minor, Si trace; Si, Al, Fe major, Ca minor, Mn, Zn trace.
GECO .32 S&W long. Unjacketed Pb bullet	Large number	None	Very small number	Small number	All the brass particles were irregularly shaped whereas the Pb-only particles were a mixture of irregular and spherical. The Pb-only particles contained some or all of Al, Ca, Cl, Cu, S, Si at minor or trace level. The few Pb, Sb particles were spherical. Unusual particles detected were Fe, Cr major, Ni minor, Mn trace; Ti major, Fe, Si minor, Al trace.
GECO .32 S&W. Unjacketed Pb bullet	None	Very small number	Large number	None	Numerous Pb, Sb particles and a few Sb only were detected. No other particle types were detected. The particles were mainly spherical. All particles contained Sn and Ti at minor or trace levels in addition to Ca, Cu, Fe, S, Si at minor or trace level.

Table 20.1 Particles on New Ammunition (Continued)

Ammunition	Pb Only	Sb Only	Pb, Sb	Brass	Observations
GECO .38 Special. Unjacketed Pb bullet	Large number	None	None	None	Numerous Pb-only particles and a few Fe particles detected. No other particle types detected. The particles were a mixture of irregular and spherical shapes. All the Pb-only particles contained Sn and Ti at minor or trace level in addition to S, Si.
GECO .38 S&W. Unjacketed Pb bullet	None	None	None	None	None of the particles contained Pb, Sb, or Ba. A large number of predominantly irregular particles were detected containing some or all of the following: Al, Ca, K, Fe, Si, Ti at major, minor or trace level, Cr, Mg at minor or trace level, Cl, Cu at trace level.
LAPUA 9 mm Luger. Cu Jacketed bullet	Small number	Very small number	Small number	Very small number	All the brass particles were irregular whereas the Pb only, Sb only, and Pb, Sb particles were all spherical. There were numerous particles containing some or all of the following: Al, Ca, Cl, Cr, Fe, Mg at major, minor or trace level, K, Ni at minor or trace level, Cu, Ti at trace level. Unusual particles detected were Fe, P, Si major, Ca minor, Cl, Cu trace; Fe, Cr, Cl major, Si trace.

Table 20.1 Particles on New Ammunition (Continued)

Ammunition	Pb Only	Sb Only	Pb, Sb	Brass	Observations
LAPUA .38 SPL. Unjacketed Pb bullet	Large number	None	Very small number	Large number	All the brass particles were irregular whereas the Pb-only particles were a mixture of irregular and spherical. The Pb, Sb particles were spherical. Several Fe particles were detected. Unusual particles detected were Zn major, S, Si minor, Ca, Cr, Fe, trace; Cr, Fe major, Si minor, S trace.
LAPUA .357 MAG. Brass case and primer cup Half Cu jacket, Pb H.P bullet	Very small number	None	Large number	Very small number	All the brass particles were irregular as were the Pb-only particles. The Pb, Sb particles were a mixture of irregular and spherical and accompanying elements were Ca, Cu, S, Si at minor or trace level.

Note: See Glossary for firearms/ammunition-related abbreviations.

showed a marked increase in loss. Again, this is predictable as the bullet travels at a considerably higher velocity (pressure) than the .380 revolver bullet and is consequently subjected to greater stress. Barrel length and rate of rifling twist may be among other contributing factors. The three sources of weight loss are erosion of the base by the hot propellant gases, engraving of the outside surface by the rifling of the barrel, and friction. It has been noted in casework that fired bullets with exposed bases frequently have powdered lead at the base area. Also noted on some occasions are embedded propellant granules or indentations caused by the granules, in the base of the bullet.

Although the weight loss may appear to be insignificant in terms of the total weight of the bullet, it is not insignificant in terms of its potential to

Table 20.2 Particles on Unloaded Ammunition

Ammunition	Pb, Sb, Ba	Sb, Ba	Ba, Ca, Si	Pb, Ba	Pb, Sb	Pb Only	Sb Only	Ba Only
GECO 9 mm Luger	37	1	None	6	24	>100	10	None
GECO .38 S&W	18	2	2	1	>100	>100	3	1

Note: See Glossary for firearms/ammunition-related abbreviations.

Casework-Related Tests

Table 20.3 Bullet Weight Loss on Firing

Bullet Type	Weight Range Before Firing (g)	Weight Range After Firing (g)	Weight Loss Range (g)	Weight Loss Range (%)	Average Weight Loss (g)	Average Weight Loss (%)
9 mm P Blazer FMJ (base enclosed)	7.4322 → 7.4741	7.4206 → 7.4630	0.0111 → 0.0125	0.1485 → 0.1678	0.0119	0.1592
9 mm P RG FMJ (base exposed)	7.5166 → 7.6288	7.4951 → 7.6033	0.0176 → 0.0284	0.2316 → 0.3770	0.0224	0.2942
.380 REV R.P Unjacketed Pb	9.3935 → 9.5412	9.3535 → 9.4936	0.0271 → 0.0476	0.2862 → 0.4989	0.0364	0.3850
W-SUPER-W .38 SPL +P Unjacketed Pb	10.2137 → 10.2679	10.1214 → 10.1962	0.0625 → 0.1051	0.6104 → 1.0277	0.0840	0.8490

Note: See Glossary for firearms/ammunition-related abbreviations.

produce a large number of discharge residue particles originating from the bullet. This work supports the proposition that the bullet makes a contribution to the discharge particle population.

Effect of Water on FDR

It has been observed that in casework involving examination of damp or wet clothing for FDR, the success rate was very low (such clothing would be dried before sampling). Possible explanations are that the particles are chemically attacked by water, that the water removes the particles by physical disturbance, for example, washed away by rain, that the water moves the particles farther into the fabric of the garment and the sampling procedure fails to recover them, or that all the cases submitted just happen to be negative. Laboratory experience and casework details make the last two options unlikely. In an attempt to clarify the situation several tests were conducted. The first test involved sampling of the firing hand immediately after firing using the same swabbing material but three different solvents, two of which had water added to them. Results are presented in Table 20.4.

Given the random nature of FDR deposition and particle recovery there is insufficient evidence to draw any conclusions from this test, although the presence of water does not appear to have a noticeable detrimental effect.

Table 20.4 Effect of Water in the Swabbing Solvent

Solvent	Particles Detected
Petroleum ether	11 × Pb, Sb, Ba; 3 × Sb, Ba; 34 x Pb, Sb; 3 × Pb, Ba
Acetone-water	58 × Pb, Sb, Ba; 46 × Pb, Sb; 4 × Pb, Ba
Acetonitrile-water	6 × Pb, Sb, Ba; 8 × Pb, Sb

The next test involved the distribution of lead in FDR between two solvents, namely, petroleum ether and water, in an attempt to determine the effect of water on the level of lead in FDR; lead was chosen because it is present in FDR at a much higher level than either antimony or barium. Separation funnels on a vibration-free surface were used for the test. Before use, both the petroleum ether and deionized water were analyzed for lead with negative result. The results are given in Table 20.5.

Again, there is insufficient evidence to draw conclusions, and the test results are difficult to explain. The lost lead from the petroleum ether layer could have been adsorbed on to the surface of the separating funnel and/or concentrated at the petroleum ether/water interface. A small amount of lead did enter the water layer in tests 2 and 3 but none in test 1. This could be due to a small proportion of the discharge residue containing a water-soluble lead compound or a small number of insoluble lead-containing particles finding their way into suspension in the water layer.

A further test involved repeatedly treating a sample with water prior to carbon coating for manual SEM/EDX examination, with a duplicate, untreated sample acting as a control. When examined, both samples had a high concentration of particles encompassing the complete range of particle types. The sample treated with water did not show any noticeable difference.

There is nothing to indicate that water has a significant chemical effect on the particles. It is likely that water, in the form of rain, would substantially decrease particle population by physical disturbance.

Table 20.5 Lead Distribution Between Layers

Sample	Test No.	Initial (ng)	48 Hours (ng)	Difference (ng)
Petroleum Ether Layer	1	2,700	1,900	800
	2	1,900	275	1,625
	3	4,025	3,750	275
Water Layer	1	None	None	None
	2	None	100	100
	3	None	50	50

Bullet Fragmentation

As a result of a terrorist attack on a motor vehicle, in which the terrorists used 7.62 mm × 39 mm caliber Yugoslavian nny 82 ammunition, the driver was shot dead. A large number of bullets struck the car, and the interior of the car and the clothing of the deceased suffered severe bullet fragmentation damage. An item of clothing worn by the deceased was examined for FDR, not as a requirement of the case but to gain background knowledge of the types of particles originating from bullet fragmentation.

Examination revealed that the sample contained both spherical and irregular particles, although the vast majority of particles were spherical. The spherical particles could originate from the considerable heat generated when a high velocity bullet strikes a hard surface, such as vehicle glass or bodywork.[196]

Numerous lead, antimony particles were detected accompanied by copper, zinc particles, iron particles, and lead-only particles. The lead, antimony; copper, zinc; and lead-only particles, probably originated from the bullets and the iron particles probably originated from the car bodywork. No unique FDR particles or other FDR particle types were detected.

If required to examine a person for FDR who had been subjected to bullet fragmentation, the presence of such large numbers of particles originating from fragmentation would make the task very difficult. In this instance no unique FDR particles were detected. However, it is possible that all types of FDR particles could be carried on the surface of the bullets, and this possibility would have to be carefully considered in this type of examination.

RPG7 Rocket Launcher

The Soviet RPG7 antitank rocket launcher using a PG7 tank rocket has been used during the terrorist campaign in Northern Ireland. It is a long weapon that sits on top of the shoulder when in use and "exhausts" to the rear of the firer. After incidents in which the RPG7 was used, the laboratory was requested to examine swabs and clothing from suspects, for discharge residue from the launcher. Because the "exhaust" from the weapon emerges a considerable distance to the rear of the firer, and the mechanism involved in its use, it was considered unlikely that residue would be present on the firer. To determine whether or not it was worthwhile examining swabs and clothing from suspects, a test was conducted and discharge residue particles from the RPG7 were examined. Discharge residues detected on the upper outer garment of the "firer" of the RPG7 are given in Table 20.6.

Discharge residue particles remaining in the launcher were also examined and the results are given in Table 20.7.

Table 20.6 Discharge Residue from RPG7 Rocket Launcher

Size μ	Shape	Major	Minor	Trace	No. Particles	Comments
3.0	Sphere	Pb	Si	Cu, Ca, Cl, Al	7	2 × Fe, S trace
3.0	Triangle	Zr, Si	Ca	Fe, K, Cu, Cl	2	From primer?
2.0 × 5.0	Oval	Pb, Si	Ca, K, Al, Fe	Cu, Cl, Ti, Mg	1	
2.0	Sphere	Cu, Pb	Sb, Fe, Zn	Al, Si, Cl, K	1	
10.0	Spherical	Pb	Ba, Cr, Fe, Ca	Cu, Si, Al	1	
3.5	Sphere	Pb, Ca	Si, Fe	Al, K, Mg, P, Cu	10	
2.0	Spherical	Sb, Sn	Cu	Fe, Si, Cl, S, Al, Mg	3	Tin present
1.5	Spherical	Pb, Si, Ca	Cl, Fe, Ti, Mg, K	Al, Cu, Zn	1	
2.5	Spherical	Pb, Ca	Ba, Si, Cl, Fe, K	Al, Mg, Cu, P, Zn	2	
1.5	Oval	Pb	Cr, Ti, Ca, Si, Zn	Cu, Cl, K, Al, Na	1	Zn > Cu
3.0	Sphere	Pb, Fe	Ca, Si, Al, Cl	Ti, Cu, K, Mg	4	
2.5	Oval	Pb, Cl	Ca, Si, Cu, Fe, K	Al, Zn, Ba, Mg	6	1 × Fe major
3.0	Double sphere	Pb, Si	Ba, Ca, Fe, Zn	Al, Mg, K, Cu, Cl	1	Zn > Cu
4.0	Spherical	Pb, Ca	Si, K, Cu, Fe	Al, Mg	21	
1.5	Sphere	Sb	Fe, Cl, Si, Al	S, K, Cu, P, Mg	3	
2.5	Oval	Pb	Ba, Si, Ca, Fe, K	Al, Zn, Cu, Mg	2	Zn > Cu
1.5	Spherical	Pb	Ca, Cr, Si, Fe	K, Cu, Zn, Al, Mg	1	
3.5	Oval	Zn	Fe	Al, K, Si	3	Zn only
3.0	Spherical	Fe	Si	Cr, Al, P, Ti	6	

The PG7 rocket is known to contain the following: black powder, a mix containing RDX explosive, hydrocarbon wax and an orange dye, PETN explosive, a mercury fulminate primer containing zirconium (see particles

Casework-Related Tests

Table 20.7 Residue in Discharged Warhead

Sample	Size μ	Shape	Major	Minor	Trace	No. Particles	Comment
Booster	3.5	Diamond	Pb	K, Ca, Co, Fe, Zn	Cu, Si, Al, P, Mn	6	Zn > Cu
Propellant	3.0	Spherical	Si, S, Pb, K, Ti	Fe, Ca, Co, Sr	Cu, Zn	2	Sr from tracer?
	7.0	Spherical	K	Si, Pb, Ca, Fe, Co	Zn	1	Zn only
Front end and outside	5.0	Rounded	Ba, Ti			5	Ti from paint?
	3.5	Globular	Si, Ti	S, K, Ca, Sr	Fe, Cu, Zn	8	Sr from tracer?
Primer	2.0	Spherical	Ba, K, S	Al, Si, Fe, Cu	Pb	14	
	3.0	Sphere	Ba, K, S, Cr	Al, Si, Ca, Cu	Zn, Pb	2	
	2.5	Spherical	Zr	Si, K, Ca	Fe, Cu	2	Zr powder?
	1.5	Sphere	Al, Si, S, Pb, K	Ba, Fe, Cu		5	
	4.0	Sphere	Co, Si, Pb	K, Ca, Fe	Cu	2	Co from propellant?
	1.5	Spherical	Pb, S, K	Al, Si	Fe, Cu	11	
	2.5	Spherical	S, Pb, Co	Al, Si, K, Ca	Fe, Cu, Zn	6	Co from propellant?
	4.5	Oblong	S, Pb, Co	Al, Si, Fe	Cu, Zn	5	Co from propellant?
	3.0	Irregular	Co		K, Ca	4	Co from propellant?
	12.0	Irregular	Si, Ba	Cu, Sr, Al, Si	K, Ca, Fe	1	Sr from tracer?
	3.0	Spherical	Hg	K, Ca, Sb, Fe	Cu	3	From primer?

in Table 20.7); an ignition powder containing barium nitrate, barium peroxide, magnesium, phenol-formaldehyde; a tracer composition containing strontium nitrate, magnesium, polyvinylchloride, phenol-formaldehyde with a yellow dye, an initial propellant charge containing NC, NG, EC, DBP, and a rocket propellant containing NC, NG, TNT (with some DNT), DPA, EC, DBP, hydrocarbon and salts of lead and cobalt. In addition the assembly contains steel, aluminum, and tin-coated copper parts, and is painted an olive drab color with black markings.

Apart from the possibility that residue was deposited on the firer in the act of firing there is also the possibility that subsequent handling of the launcher could yield distinctive residue. It is worthwhile examining the suspect for such residue.

Discharge Residue from Black Powder Ammunition

In previous casework in which the majority of the FDR particles contained potassium and sulfur, frequently at high levels, it was thought that the ammunition responsible probably contained black powder. In most of the cases the type of ammunition was not known, whereas in others the sampling and analysis of the residue from the interior of the spent cartridge cases confirmed the presence of black powder.

This posed the question "do potassium and sulfur always occur, often at high level in discharge residue particles from ammunition loaded with black powder?" In other words, from the presence and levels of potassium and sulfur can it be accurately predicted when black powder is used?

A selection of old ammunition was tested to confirm that the propellant was black powder. Results of a representative selection of SEM/EDX analysis of the undischarged black powder are presented in Table 20.8.

It is interesting to note the presence of lead, antimony, and mercury in some of the analyses. The mercury is almost certainly from the primer whereas the lead and antimony could originate from two sources: the base of the bullet or the primer. However, if they originate from the bullet it would be expected that they would occur together and that the lead level would be significantly greater than the antimony level. This suggests that the lead and antimony also originate from the primer.

Discharge residue particles from black powder ammunition were then examined. Table 20.9 gives representative results.

As antimony sulfide is widely used in primer compositions, sulfur is frequently present in discharge residue particles and can occur at major, minor, or trace level (see Table 19.5). Consequently the occurrence of sulfur at major level is not an accurate indicator of the use of black powder. The particles should be considered as a group and it is clear that the frequent occurrence of both potassium and sulfur at high level is strongly indicative of black powder. However, as can be seen from Table 20.9 the use of black powder does not necessarily yield overall high levels of potassium. Potassium does not normally occur at major level in FDR particles (see Table 19.5) and its presence at major level in any of the particles suggests the use of black powder.

Casework-Related Tests

Table 20.8 Analysis of Unburned Black Powder

Ammunition	Major	Minor	Trace	Comment
UMC 32-20	K, Pb, S	—	—	
	Pb, S, K	—	Si	Numerous
	Pb, S	K	Si	
	Pb, S, K	—	Si, Zn	
	Pb, S	—	K, Si, Zn	
	Hg, K	—	Si, Cu, Zn	
	K, Hg	—	Cu, Si	
	Pb, S	—	K, Si	
RWS .320	K	—	Pb, S, Cu	
	K	S	Pb, Si	
	K	—	Pb, Cu	
	S, K	—	Pb, Si	
	Pb, S, K	—	Si, Cu	Numerous
No head stamp 297/230 Morris	K	—	S	
	S	—	K	Numerous
	K, S	—	Cu	
RWS .380	Pb, S, K	—	Si	Numerous
	S, Sb, K	—	Si	
	Pb, S	K	Si	
RWS .450	S, Sb, K	—	Si	
	Pb, S	K	Si	
	Pb, S, K	—	Si	Numerous
	S, Sb, K	—	Si	
	Pb, S, K	—	Cu, Zn, Si	
Eley London .450	K, Pb, S	—	Si	
	Pb, S, K	—	Si	Numerous
	K, Sb	—	S, Si	
	K, S	Pb	Si	
	S, Sb	K	Si	

Note: See Glossary for firearms/ammunition-related abbreviations.

A potential problem arises whenever black powder ammunition is used in close range shooting, in that the particulate matter deposited in the vicinity of the bullet hole is nondescript and does not resemble smokeless propellant. Consequently its significance may not be realized and it is also difficult

Table 20.9 Discharge Particles from Black Powder Ammunition

Ammunition	Size μ	Shape	Major	Minor	Trace	Comments
UMC 32-20	5.0	Oval	Pb, S	—	Si	Numerous No Hg detected (Overall high S, trace K)
	4.0	Kidney	Pb, S, Ba, Sb	—	Si, Fe, Cu	
	3.5	Sphere	Pb, S	—	Si, Cu, Sb	
	2.0	Oval	Pb, S, Si, Ba	Ca	K, Fe, Cu	
	2.0	Oval	Sb, Si	Pb, S	Ba, Cu, Fe	
RWS .320	8.0	Irregular	Pb, S	—	Si, Cu	Numerous (Overall high K and S)
	4.0	Oval	K, S	—	Cu	
	3.0	Oval	Pb, S, K	Cl, Si, Ca	Fe, Cu, Ba	
	5.0	Sphere	Sb, Pb, S, Fe	Cl, K	Si, Cu	
No head stamp 297/230 Morris	1.5	Oval	Pb, S	—	Si, Ti, Fe, Ca, Cu	Numerous (Overall high S, low K)
	4.0	Irregular	Fe	Cr		
	3.0	Oval	Pb, S	—	Si, Sb, Ti, K, Fe	
	3.0	Sphere	Fe, Pb, S, K	Ca	Ni	
	8.0	Oval	K, Si, Ba, Ca, Pb, S	Fe	Si, Ti, Cu	
RWS .380	7.0	Oval	Pb, S	—	Si, Ca, K, Fe, Ti	Numerous (Overall high S, low K)
	5.0	Spherical	Pb, S, K, Ca	—	Si, Ti, Fe, Cu	
	3.5	Oval	Pb, S, Ba	Sb, K, Si	Fe, Cu	
	7.0	Kidney	Ba, Si, Ca	—	K, Fe	
	3.0	Triangle	Pb, S	—	K, Fe, Cu	
RWS .450	10.0	Irregular	K, S	—	—	Numerous (Overall high K and S)
	3.0	Spherical	Pb, S, K, Cl	Si	Fe, Cu	
	5.0	Oval	Cl, K	—	Si	
	12.0	Oval	Sb, Ba, Pb, S	Si	K, Cu, Fe	
	12.0	Irregular	S, Pb, K, Fe	Cl	Cu, Zn	
Eley London .450	4.5	Spherical	K, S	—	—	Numerous (Overall high K and S)
	2.0	Oval	S, Pb, Sb	Ba	Fe, Cu, Cl	
	8.0	Irregular	Pb, S, Cl	K, Ca, Si	Fe, Cu	
	2.5	Oval	S, Pb, K	—	Cl, Si	
	2.0	Oval	S, Sb, Ba	Cl	K, Cu, Fe	

Note: See Glossary for firearms/ammunition-related abbreviations.

to see on dark surfaces. If found it should be examined for potassium and sulfur for confirmation of black powder.

Firearm Coatings

The surface coating of a random selection of firearms was examined and the results are given in Table 20.10. The elements detected in Table 20.10 do not fully reflect the wide range of elements mentioned in Chapter 15.

It is possible that such coatings could, on rare occasions, make a contribution to the elemental content of some of the particles, or could be deposited directly on to the hands from handling the firearm. Surface coatings have the potential to make a contribution, particularly from the cylinder gap area of revolvers and from the muzzle area of any type of firearm. These are the "exterior" areas subjected to the hot propellant gases, which may erode the surface coating. Mixtures containing selenium are used to repair surface coatings of firearms, and particles containing selenium are occasionally encountered in casework. Such particles can provide useful additional evidence. Homemade firearms are frequently painted black using household paints. Such coatings can flake and leave paint flakes on the hands or clothing (particularly pocket interiors), and this can provide very useful evidence.

Homogeneity of Propellants

To investigate the feasibility of conducting chemical comparisons between propellants detected in casework and suspect ammunition, it was decided to determine variations from granule to granule in a single round of ammunition and then to compare burned (discharged) and unburned propellants, to determine what, if any, difference was caused by the discharge process.

Analysis of 20 propellant granules from a single round of ammunition gave the following results. Diphenylamine and three dialkylphthalates were detected in all 20 granules, whereas NG was detected in only 2 of the granules. Peak area ratios of DPA to each phthalate varied widely for each granule. This, and the fact that NG was detected in only 2 out of 20 granules in a single round of this particular ammunition, shows a considerable compositional difference between granules of propellant. It is a possibility that this is a blended propellant. Further work would need to be done in this area to determine granule to granule and batch to batch variations in composition, in order to determine the feasibility of chemical comparisons in each particular instance.

A further test explored compositional differences between fired and unfired propellant granules. Results are given in Table 20.11.

Table 20.10 Firearm Surface Coatings

Firearm	Surface Appearance	Major	Minor	Trace
Steyr Grand Rapide .308 Win	Gray, matt	Fe, Mn, P	Ca, S	Si
FAL Rifle 7.62 × 51 mm	Black, gloss	Si, Cl	P, Ca, Si, Ba	S, K, Fe, Ti, Na
Brno Mod 38 × 22 Rimfire	Black, gloss	Fe	—	Mn, S, Si
Sig Manurhin .243 Win	Gray, matt	Mn, Fe, Ca	K	P, Si, S, Cu
FNC Rifle .223 Rem	Black, matt	Fe	—	Mn
H&K MP5 SMG 9 mmP	Blue, matt	Si	—	Mg
Webley Vulcan Air Rifle .22	Black, gloss	Fe	—	Mn, S
Beretta O/U Shotgun 12 G	Black, gloss	Fe	—	Mn, Cr, Ca, Si
Beretta 302 S/A Shotgun 12 G	Black, gloss	S	Ni	Fe
Colt AR-15 Rifle .223 Rem	Gray, matt	Fe, Mn, P	Ca, Cl, K, S	—
Sterling SMG 9 mmP	Black, gloss	Si, Fe	P, S, Mn, Cl	K
MI Garand Rifle .30-06	Gray, matt	Fe	Zn	S, Si
Baikal S/B Shotgun 12 G	Black, gloss	Fe	Cr, K	S, Cl, Si
Aya D/B Shotgun 12 G	Black, gloss	Fe	Mn	K
Gardone O/U Shotgun 12G	Black, gloss	Fe	Mn, Si	S, Cl
Steyr 1904 Rifle 7.9 mm	Black, gloss	Fe, Cl, K	Si	Mn, Cu, S
Walther Pistol .380 ACP	Black, gloss	Fe	Mn, K, S, Si	Ca, Cl
S&W Mod .59 Pistol 9 mmP	Black, gloss	Al	S, Cl, Si	K, Ca, Fe, Ni, Zn
S&W 15-4 Revolver .38 SPL	Black, gloss	Fe	—	S
Browning Pistol 9 mmP	Black, matt	Fe	—	K, Cl, Ca
Ruger Speed Six .357 Mag	Black, gloss	Fe	—	Cr
Ingram SMG 9 mmP	Black, matt	P, Zn	Fe, Sb, Ca, K	Si, S, Cl, Cu
Sussex Armoury Replica	Black, matt	Si, Ba, S	Mg, Al	Fe, Cu
Webley & Scott .38 S&W	Black, matt	Fe	S, K, Cl, Ca	Si, Cu
Colt Pistol .45 ACP	Black, matt	Fe	S, Al	Cr, Mn, Cu, Cl, K
MI Carbine .45 ACP	Black, matt	Si, S, Fe	Ca, Cl, K, Al	Ni, Mn, Cu
Franchi S/A Shotgun 12 G	Black, gloss	Fe	Cl, S	K, Cu, Si
Webley & Scott S/B Shotgun 12 G	Black, gloss	Fe	—	Si, S, Cl, K, Ca, Cu
Lee Enfield Rifle .303	Black, matt	Fe, Ca	K, S, Cl, Si	Cu, Zn
AR-180 Rifle .233 Rem	Green, matt	P, Mn, Fe	—	Ca, Cu

Note: See Glossary for firearms/ammunition-related abbreviations.

Table 20.11 Comparison of Fired and Unfired Propellants

							Phthalates			
Sample	NG	EC	DPA	nDPA 1	nDPA 2	Tri-butyl-phos.	P1	P2	P3	P4
A Fired	Minor	Major	ND	ND	ND	ND	Trace	Major	ND	Minor
A Unfired	Major	Major	Trace	ND	ND	ND	Trace	Trace	Minor	Trace
B Fired	Minor	Trace	Major	Trace	Trace	ND	ND	Trace	Trace	Trace
B Unfired	Minor	Trace	Major	Trace	Trace	ND	ND	Major	Trace	Minor
C Fired	Trace	ND	Trace	ND	ND	ND	ND	ND	ND	ND
C Unfired	Minor	Trace	Major	Trace	Trace	ND	ND	Minor	Minor	Minor
D Fired	Minor	Trace	Major	Trace	Trace	ND	ND	Minor	Trace	Trace
D Unfired	Minor	ND	Major	Trace	Trace	ND	ND	Minor	ND	Minor
E Fired	Minor	Major	Minor	ND	ND	ND	ND	Major	ND	Trace
E Unfired	Minor	Major	Minor	ND	ND	ND	ND	Major	ND	Trace
F Fired	Minor	Minor	Major	Trace	Trace	ND	ND	Minor	ND	Trace
F Unfired	Minor	Trace	Major	Trace	Trace	ND	ND	Major	ND	Major
G Fired	Minor	Trace	Major	Trace	Trace	ND	ND	Major	ND	Minor
G Unfired	Minor	Trace	Major	Trace	Trace	ND	ND	Minor	ND	Trace
H Fired	Minor	Minor	Major	Trace	Trace	ND	ND	Major	ND	Trace
H Unfired	Minor	Trace	Major	Trace	Trace	ND	ND	Major	ND	Trace
I Fired	Major	Major	Trace	ND	ND	ND	ND	Major	ND	Trace
I Unfired	Major	Major	Trace	ND	ND	Trace	ND	Major	ND	Trace
J Fired	Minor	Major	ND	ND	ND	Trace	ND	Major	ND	Trace
J Unfired	Minor	Major	ND	ND	ND	ND	ND	Minor	ND	Trace

DPA, diphenylamine; EC, ethylcentralite; ND, none detected; NG, nitroglycerine.

Some ingredients such as EC may be present as a surface coating, as opposed to an integral component, and may be blown or burned off during discharge. It is clear from this work that whatever the reasons for compositional differences, whether qualitative or quantitative, any interpretation based on compositional data needs to be approached with caution, particularly when only a small quantity is available for comparison. Any conclusive interpretations would need to be supported by a massive amount of background data.

Bullet Hole Perimeters

The sodium rhodizonate test for lead[197] is routinely used in many forensic laboratories for confirmation of bullet damage and range of fire determinations. In a number of cases the test failed to indicate the presence of lead on the perimeter of holes that had a distinct bullet wipe. The bullets involved in these cases were all copper jacketed (FMJ). As the test is routinely used in this laboratory, the negative findings from known bullet holes caused concern. It was decided to instigate a project to assess the reliability of the sodium rhodizonate test for lead, and the validity of lead as an indicator of bullet damage. Also investigated was the bullet wipe pattern for shots fired from different angles, and the dependence of close range residue patterns on the ammunition used.

The first test involved single shot firings at a fixed distance, using the same gun and ammunition type, but varying the angle of the target and the angle of the firer to the target. A "straight on" (0°) shot will produce a uniform circular hole and wipe, whereas a shot fired from an angle will produce an elongated bullet wipe and a somewhat irregular hole. One of the questions to be tested was "does the size and position of the elongated wipe reliably indicate the angle of fire?"

Test results revealed that 62 of the 63 bullet hole perimeters gave a positive rhodizonate test for lead and it was concluded that the size of wipe produced by the bullet increases as the angle of fire increases, for example, a shot fired from 75° (left or right) will produce a larger wipe than a shot fired from 30° (left or right). Shots fired straight on at the target only produced an elongated wipe whenever the target was tilted.

There was a definite tendency for the size and position of the wipe to be reproducible for repeated firings under identical conditions, but in a few instances it varied markedly, without apparent reason. As the size and position of the wipe depends on the relative positions and attitudes of the firer and target, any conclusions about the direction of fire need to be very carefully considered.

The second test involved a series of single-shot close range firings using a revolver, a pistol, and a rifle, but varying the ammunition used. The objective was to determine the influence of the type of firearm and the type of ammunition on the muzzle blast residue pattern deposited on the target in close range shootings.

Results indicated that in each case the diameter and density of the unburned propellant patterns were similar using the same gun but different ammunition. There were variations in the soot (blackening) deposits with different ammunition. Contact shots were very similar irrespective of the ammunition. All gave positive rhodizonate tests for lead.

Casework-Related Tests

Although it is always desirable to use the actual gun and the same ammunition type to do range tests for comparison with casework items, if the ammunition type is unknown and the firearm type is known, it is still possible to give a reasonable estimate of range. If both the gun and ammunition type are unknown, that is, at one extreme it could be a low power handgun and at the other it could be a high power rifle, then in these instances it is possible to state only that there is evidence of a close range shooting, give the upper limits for a handgun and a rifle, and then give a rough estimate for each couched in terms such as "not more than" and "not less than."

A final test was conducted to determine the reliability of the sodium rhodizonate test as an indicator of bullet damage. This revealed that approximately 99% of the ammunition used in the pistols and revolvers gave positive rhodizonate tests on the perimeter of the bullet hole and approximately 94% of the ammunition used in the rifles gave positive rhodizonate tests. These results indicate that the sodium rhodizonate test for lead is reliable and that lead is a good indicator of bullet damage and close range shootings. The results are better than those experienced in casework because, in casework, many bullet hole perimeters are bloodstained and the blood could disturb the perimeter residues and have a masking effect, thereby hindering the removal of residue for testing. Despite this the test is effective for the vast majority of cases. The lower success rate with rifles is difficult to explain, but may well be a result of some of the lightly adhering residue on the bullet surface being lost due to the higher velocity (wind disturbance) before the bullet strikes the target.

Alternative tests for the identification of bullet holes and testing for close range shooting will need to be devised as the use of lead-free ammunition increases.[198]

Tests were also conducted to determine if it was possible to identify the bullet jacket material from examination of the bullet hole perimeter. The ammunition used is given in Table 20.12 and the test results are presented in Table 20.13.

The residue on the surface of a discharged bullet appears to originate from the base of the bullet itself, from the primer, and from inorganic additives to the propellant. Firings numbered 8, 21, 34, and 35 had lead-free primers yet lead was detected on the perimeter of the bullet holes. Ammunition with barium-free primers gave barium on the perimeter.

Only one of the two nickel-jacketed bullets, number 43, gave nickel on the perimeter of the bullet hole. Nickel was frequently detected from non-nickel-coated bullets. This is a surprising result which demonstrates that the presence of nickel cannot be used to identify the use of a nickel-jacketed bullet. The origin of the nickel is unknown but it may have originated from the primer cup coating.

Table 20.12 Bullet Hole Perimeter Test Ammunition

Test No.	Ammunition		Primer
1	9 mmK Hirtenberg	Ni Jkt FMJ	Pb, Ba
2	9 mmK Sako	Cu Jkt FMJ	Pb, Sb
3	9 mmK W-W	Cu Jkt FMJ	Pb, Sb, Ba
4	9 mmK Federal	Cu Jkt FMJ	Pb, Sb, Ba
5	9 mmP VPT42	Cu Jkt FMJ	Pb, Sb, Ba
6	9 mmP VPT43	Cu Jkt FMJ	Pb, Sb, Ba
7	9 mmP VPT44	Cu Jkt FMJ	Pb, Sb, Ba
8	9 mmP 11 52	Cu Jkt FMJ	Sb, Hg
9	9 mmP K52	Cu Jkt FMJ	Pb, Sb, Ba
10	9 mmP S044	Cu Jkt FMJ	Pb, Sb, Ba
11	9 mmP GECO 80-59	Cu Jkt FMJ	Pb, Sb, Ba
12	9 mmP Norma	Cu Jkt FMJ	Pb, Sb, Ba
13	9 mmP REM-UMC	Cu Jkt FMJ	Pb, Sb, Ba
14	9 mmP RG55	Cu Jkt FMJ	Pb, Sb, Hg
15	9 mmP D143	Cu Jkt FMJ	Pb, Ba
16	9 mmP RG56	Cu Jkt FMJ	Pb, Sb, Hg
17	9 mmP RG57	Cu Jkt FMJ	Pb, Sb, Hg
18	9 mmP WRA	Cu Jkt FMJ	Pb, Sb, Ba
19	.45 ACP R-P	Cu Jkt FMJ	Pb, Sb, Ba
20	.45 ACP W-W	Cu Jkt FMJ	Pb, Sb, Ba
21	.45 ACP SF57	Cu Jkt FMJ	Sb, Hg
22	.45 ACP WRA. Co	Cu Jkt FMJ	Pb, Sb, Ba, Hg
23	.303 R↑L49	Cu Jkt FMJ	Pb, Sb, Hg
24	7.62 NATO RG70	Cu Jkt FMJ	Pb, Sb, Ba
25	.223 HP	Cu Jkt FMJ	Pb, Sb, Ba
26	.223 Norma	Cu Jkt FMJ	Pb, Sb, Ba
27	.223 IV170	Cu Jkt FMJ	Pb, Sb
28	.223 RA69	Cu Jkt FMJ	Pb, Sb, Ba
29	.223 RA65	Cu Jkt FMJ	Pb, Sb, Ba
30	.30MI Norma	Steel Jkt JSP	Pb, Sb, Ba
31	.30MI R-P	Steel Jkt JSP	Pb, Sb, Ba
32	.30MI W-W	Cu Jkt FMJ	Pb, Sb, Ba
33	.30MI W-W	Cu Jkt FMJ	Pb, Sb
34	.30MI VE-F	Cu Jkt FMJ	Sb, Hg
35	.30MI VE-N	Cu Jkt FMJ	Sb, Hg
36	.455 Dominion	Pb unjacketed	Pb, Sb, Ba, Hg

Casework-Related Tests

Table 20.12 Bullet Hole Perimeter Test Ammunition (Continued)

Test No.	Ammunition		Primer
37	.455 Kynoch	Pb unjacketed	Pb, Sb, Ba
38	.455 K62	Cu Jkt FMJ	Pb, Sb, Ba, Hg
39	.357 W-W	Cu Jkt JHP	Pb, Sb, Ba
40	.357 R-P	Cu Jkt JSP	Pb, Sb
41	.357 W-W	Pb SWC	Pb, Sb, Ba
42	.357 R-P	Pb SWC	Pb, Sb, Ba
43	.38 S&W R↑L39	Ni Jkt FMJ	Pb, Sb, Hg
44	.38 S&W Norma	Pb unjacketed	Pb, Sb, Ba, Hg
45	.38 S&W REM-UMC	Pb unjacketed	Pb, Sb, Ba, Hg
46	.38 S&W Browning	Pb unjacketed	Pb, Sb, Ba
47	.38 S&W GECO	Pb unjacketed	Pb, Sb, Ba

Note: See Glossary for firearms/ammunition-related abbreviations.

It is interesting to note that in all tests in which mercury was present in the primer, it was detected on the perimeter of the bullet hole. The unjacketed lead bullets all gave a large quantity of lead on the perimeter, although this was not confined to unjacketed bullets. The copper results were similarly confusing.

Overall, the possibility of determining the bullet jacket material from the residue around the bullet hole does not appear to be feasible using FAAS. However, FAAS reliably detects elements associated with firearm discharge on the perimeter of the bullet hole and is a very useful method for confirming bullet damage.

Persistence

An obvious trend over a 26-year period of the terrorist campaign is the decreasing percentage of Northern Ireland casework that is positive for FDR. During this period substantial improvements have been made in the efficiency of sampling and in the sensitivity of the detection techniques. Despite this, the downward trend continued. Our success rate decreased from approximately 35% at the start of the terrorist campaign in 1969 to about 6% (excluding suicides and dead suspects) in 1995. The reason for the decreasing success rate is not the detection system but rather the careful planning of terrorist incidents and the precautions terrorists take to prevent leaving any type of forensic evidence at a scene or on their persons. Coupled with this is the unfavorable behavior of FDR particles once they are deposited on a suspect. The particles are small and lightly adhering and, as such, can become

Table 20.13 Elemental Levels (ng) on Perimeter of Bullet Hole

Sample No./Jkt Material	Pb	Sb	Ba	Cu	Ni	Hg	Comments
1 Ni	2,600	None	160	600	None	None	No Ni, low Pb and Ba
2 Cu	>10,000	None	500	1,150	None	None	No Sb. Ba present
3 Cu	>10,000	38	1,690	1,500	3,700	None	High Ni
4 Cu	>10,000	None	None	None	None	None	Pb only detected
5 Cu	8,300	None	850	>5,000	None	None	No Sb
6 Cu	>10,000	None	560	4,075	2,850	None	No Sb, high Ni
7 Cu	9,350	30	410	4,650	None	None	Pb and Ba present
8 Cu	9,000	None	650	>5,000	None	63	No Sb. Ba present
9 Cu	3,950	None	520	>5,000	None	None	No Sb
10 Cu	8,875	30	520	4,750	None	None	
11 Cu	4,400	20	1,460	2,200	None	None	
12 Cu	>10,000	46	670	2,600	None	None	
13 Cu	>10,000	36	700	2,550	None	None	
14 Cu	>10,000	33	540	4,175	2,000	160	Ba present, high Ni
15 Cu	>10,000	None	830	>5,000	1,900	None	High Ni
16 Cu	9,500	41	580	4,100	None	176	Ba present
17 Cu	3,850	66	130	3,100	None	286	Ba present
18 Cu	3,830	75	400	>5,000	None	None	
19 Cu	>10,000	167	2,000	>5,000	1,725	None	High Ni
20 Cu	>10,000	137	1,470	3,275	None	None	
21 Cu	5,150	102	650	4,890	2,300	>500	Pb and Ba present, high Ni
22 Cu	9,220	None	>2,000	>5,000	None	>500	No Sb
23 Cu	>10,000	52	None	>5,000	None	>500	
24 Cu	>10,000	None	390	>5,000	None	None	No Sb
25 Cu	2,150	None	160	>5,000	None	None	No Sb, low Pb and Ba
26 Cu	5,450	None	450	2,490	None	None	No Sb
27 Cu	5,000	None	740	1,850	2,100	None	No Sb, Ba present, high Ni
28 Cu	3,375	29	450	1,400	None	None	
29 Cu	4,900	None	>2,000	4,750	2,250	None	No Sb, high Ni
30 Steel	2,950	None	1,270	>5,000	2,150	None	No Sb, high Ni
31 Steel	>10,000	55	1,050	>5,000	2,150	None	High Ni
32 Cu	8,800	30	>2,000	>5,000	None	None	
33 Cu	5,900	32	720	>5,000	None	None	Ba present

Casework-Related Tests

Table 20.13 Elemental Levels (ng) on Perimeter of Bullet Hole (Continued)

Sample No./Jkt Material	Pb	Sb	Ba	Cu	Ni	Hg	Comments
34 Cu	4,700	>200	900	>5,000	None	155	Pb and Ba present
35 Cu	850	>200	550	>5,000	None	>500	Pb and Ba present
36 Pb	>10,000	>200	>2,000	3,950	None	>500	
37 Pb	>10,000	>200	>2,000	3,325	None	>500	
38 Cu	>10,000	>200	>2,000	>5,000	2,850	None	High Ni
39 Cu	>10,000	None	240	2,200	None	None	No Sb
40 Cu	7,300	None	200	1,100	1,600	None	No Sb, Ba present, high Ni
41 Pb	>10,000	>200	520	None	None	None	
42 Pb	>10,000	None	835	None	3,675	None	No Sb, high Ni
43 Ni	>10,000	None	390	>5,000	3,250	>500	No Sb, Ba present, high Ni, Cu
44 Pb	>10,000	>200	1,830	>5,000	2,850	>500	High Ni
45 Pb	>10,000	174	1,770	3,300	4,750	>500	High Ni
46 Pb	>10,000	174	700	900	None	None	
47 Pb	>10,000	130	1,370	1,675	None	None	

airborne again and be transferred from surface to surface by physical contact. They are lost rapidly from the hands, an order of magnitude in the first hour, and consequently the detection of residue on the hands suggests very recent contact (excluding suicides and dead suspects). They persist longer on clothing surfaces, the length of time depending on the nature of the material and the extent of physical disturbance of the garment. The particles are chemically stable. This was confirmed by an experiment involving an FDR-contaminated garment which was packaged, sealed, and stored for 2 years. FDR particles were readily detected on the garment surface after the lengthy storage period.

Another persistence experiment involved prompt sampling of the firing hand after firing. Numerous FDR particles were detected on the firing hand. The experiment was repeated but the firer was allowed to dry wipe his hands on tissue, in an effort to remove any FDR particles, prior to sampling. Very few particles were detected and it was concluded that FDR particles can easily be removed from the hands, even by dry rubbing.

Statistics gathered from 15 years of casework results gave the following persistence data. Figure 20.1 illustrates the situation for suspects whose hands, face, and head hair was sampled for FDR, resulting in the detection of particles on all or some of the samples. *Suspects are rarely apprehended*

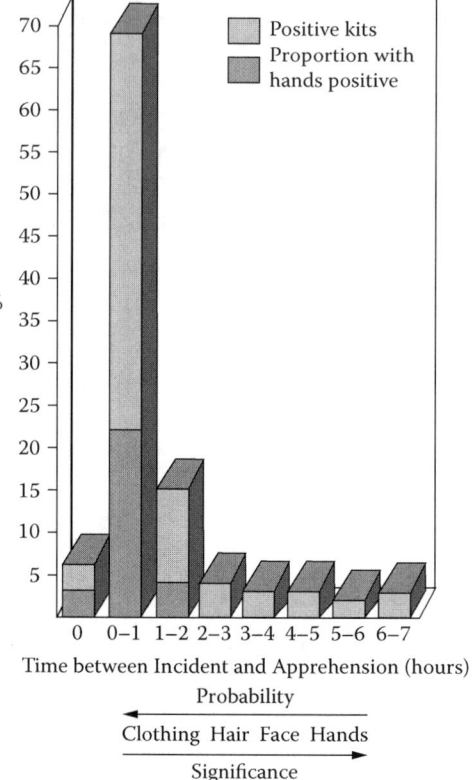

Figure 20.1 Persistence of FDR.

immediately. The data are based on 410 positive swab kits and exclude suicides and dead suspects.

It is difficult to produce valid persistence data for clothing other than to say that it is our most fruitful sampling area (pocket interiors in particular). As stated previously, persistence on clothing depends on the nature of the material and the degree of physical disturbance the garment suffers. FDR will remain indefinitely on clothing if the clothing is undisturbed.

Suspects have been known either to abandon or to destroy clothing worn at the time of the incident and change into "clean" clothing. Consequently it is often not known for certain if the clothing submitted to the laboratory was the clothing worn during the incident.

FDR has been detected on the clothing of a suspect up to 6 days after an incident, but the history of the clothing was not known; consequently the residue could have been deposited since the original incident. On the other hand residue found on a garment could have originated from an incident prior to the incident under investigation. This highlights the problems encountered when interpreting positive results on clothing. Residue detected on the

Casework-Related Tests

hands, face, or head hair, because of known persistence, can be assumed to be recently deposited whereas on clothing (particularly pocket interiors) it is difficult to link it to a specific shooting incident. However, it is still valuable evidence requiring an explanation from the suspect.

FDR is not likely to be found on the hands if the time between the incident and apprehension exceeds 2 hours (the suspect's hands must be protected immediately after the suspect is apprehended). The police are instructed not to sample the hands if the 2 hours are exceeded but to take the face and head hair samples as normal. FDR has been detected on the face up to 5 hours and on the head hair up to 7 hours after an incident.

Antimony-Free Primers

As mentioned previously it was noted in casework that Yugoslavian 7.62 × 39 mm caliber nny 82 ammunition produces discharge particles containing barium, despite the fact that there is no barium in the primer, although it is present in the propellant. This indicates that the propellant can make a contribution to the elemental content of the discharge particles. In some incidents involving the use of ammunition with antimony-free primers, discharge particles containing antimony have been detected on suspects. The possibilities are that the antimony originated from the bullet, that the particles were due to contamination of the gun or ammunition from some previous firing, that the particles originated from some other incident in which the suspects were involved, or that the suspects had been exposed to contamination between apprehension and sampling.

To clarify the situation it was decided to investigate the possibility that the antimony originated from the bullet. Discharge residue particles originating from ammunition with antimony-free primers and antimony-hardened bullets were examined for the presence of antimony. Results are given in Table 20.14.

Analysis of a Baton Round

As a consequence of a case in which the suspect alleged that the FDR on his person originated from contact with the inside of a police vehicle, from which baton guns had previously been fired, it was necessary to conduct a detailed examination of the baton round and the crime ammunition.

Antimony was detected in the residue particles on the suspect and it was known that the baton round does not have antimony in the primer. However, it was required to prove that there was no antimony in any part of a baton round. Analysis of the baton round revealed that the cartridge case was

Table 20.14 Discharge Particles from Ammunition with Antimony-Free Primers

Firearm	Ammunition Head Stamp	Ammunition Details	Pb, Sb, Ba	Ba, Ca, Si	Pb, Ba	Pb, Sb	Ba Only	Pb Only
.22LR Walther pistol	U	Unjacketed. Brass wash on bullet. Pb-only primer	ND	ND	ND	20	ND	36
9 mmP Star pistol	D144	FMJ. Cu washed Fe jacket. Pb, Ba primer	7	19	15	2	ND	15
.30 M1 Winchester carbine	LC68	FMJ. Cu washed Fe jacket. Pb, Ba primer	18	ND	12	2	6	22

ND = not detected

aluminum with trace amounts of iron and silver, and it was painted black with a white band. Analysis of the black paint showed the presence of aluminum, silver, bromine, chlorine, chromium, iron, potassium, nickel, sulfur whereas the white paint revealed titanium only. The discharged primer residue gave lead and barium at major level, aluminum and silicon at minor level, and tin and copper at trace level. The primer cup was tin with a trace of copper and silver, and the red lacquer on the exterior surface of the cup contained silicon and tin. The black powder propellant was housed in a plastic casing on which was painted a green dot. Analysis of the propellant revealed potassium and sulfur at major level, and silicon and iron at trace level, and the green paint gave lead, sulfur, potassium at major level, chromium, chlorine at minor level, and barium, titanium, iron at trace level. The plastic baton itself had chlorine, iron, barium at major level and calcium, silicon at trace level on its outside surface, and chlorine at major level and calcium at trace level on its inside surface. The end cap was painted cream and analysis showed the presence of sulfur, silicon, calcium at major level with iron, potassium, silver, aluminum, and bromine at trace level.

No antimony was detected anywhere in the baton round. Test firing the baton gun and sampling the firer also failed to reveal the presence of antimony in the discharge particles. Tin was present in some of the discharge particles.

The crime ammunition was 7.62 × 39 mm caliber Yugoslavian nny 82. Analysis of the components gave the following results.

Propellant: Single based with DPA, EC, a phthalate plasticizer, camphor
  Unburned   Si, Pb at major level, Ca, Cu, S at minor level, Ba, Fe, K at trace level
  Burned   K and S at major level
Cartridge case and primer cup: Cu with a trace of Zn
Discharge primer residue (elements listed in descending order):
  (inside of primer cup)   Sb, Sn, S, Cl, Hg, Cu, Zn, Fe, K, Ca
  (inside of cartridge case)   Sb, Sn, K, Cl, Cu, S, Zn, Fe, Hg
Bullet jacket: Cu at major level with Zn at minor level and Fe, Cl at trace level
Bullet core: Pb with trace levels of Sb, Si
Lacquer (primer):   Pb at major level, Ti, Cr, Si at minor level, Fe, Mn, Cu, Cl, K, Ca at trace level
Black sealant between bullet and cartridge case: Cu, Pb, Cl at major level, Zn, S at minor level, Si, Sb, Fe, K, Ca at trace level

It is interesting to note that the primer appears to be based on mercury fulminate, antimony sulfide, and potassium chlorate, that is, mercuric and corrosive, and that the ammunition was manufactured in 1982.

There is no lead or barium in the primer yet discharge particles from this ammunition frequently contain lead, antimony, and barium. The lead and barium must come from other components in the ammunition (bullet core/propellant) and/or from contamination in the firearm. Tin was also frequently present in the discharge particles and originates from the tinfoil disc used to seal the primer cup in mercury fulminate primers.

The absence of antimony in the baton round discharge particles proved that the residue on the suspect did not originate from this source.

References

196. J. S. Wallace, "Bullet Strike Flash," *AFTE Journal* 20, no. 3 (July 1988).
197. J. H. Dillon, "Sodium Rhodizonate Test: A Chemically Specific Test for Lead in Gunshot Residues," *AFTE Journal* 22, no. 3 (October 1990): 251.
198. R. Beijer, "Experiences with Zincon, a Useful Reagent for the Determination of Firing Range with Respect to Leadfree Ammunition," *Journal of Forensic Sciences* 39, no. 4 (July 1994): 981.

Analysis of Ammunition

21

Introduction

From the beginning of the terrorist campaign in 1969 to the end of April 1994 a total of 16,381 firearms, 82,168 spent cartridge cases, and 1,667,115 rounds of ammunition have been recovered by the security forces, along with numerous miscellaneous firearms-related items. The recovered firearms consist of 905 machine guns, 630 carbines, 4,871 rifles, 3,816 pistols, 3,414 revolvers, 2,196 shotguns, and 549 miscellaneous firearms. Calibers range from .22" up to and including 50"/12.7 mm. During this period there were 10,995 shooting incidents and analysis of some of the recovered ammunition is discussed in this section.

Primer Types

The interior of the spent cartridge case is routinely examined whenever FDR is detected on a suspect, to determine the type of primer involved in the incident. Figure 21.1 and Figure 21.2 illustrate primer types as determined by FAAS and SEM/EDX examination, respectively.

Propellant Analysis

The samples were granules of propellant recovered from materials shot at close range, mainly clothing from injured persons. Consequently, in the vast majority of instances the type of ammunition is not known. It must be noted that granules found may not necessarily be representative of the bulk, as composition can vary from granule to granule, and also that these are "discharged" propellant granules and in the act of discharge surface coatings can be blown or burned off.[199]

Table 21.1 and Table 21.2 give the propellant compositions detected over a 4-year period (1990 to 1993). Naphthalene and tributylphosphate are not mentioned in the literature on propellants.

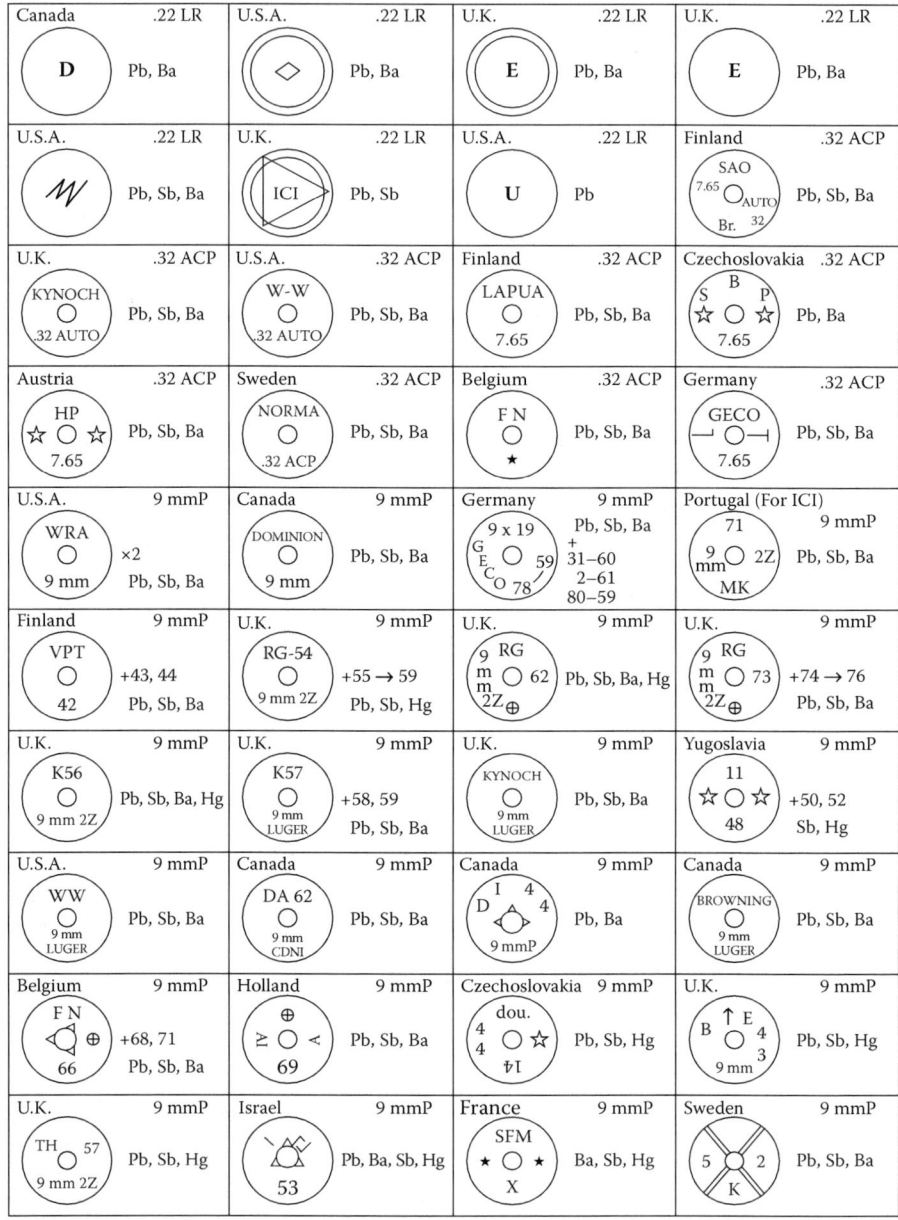

Figure 21.1a FAAS analysis of spent cartridges.

Analysis of Ammunition

Country	Caliber	Headstamp	Components
Austria	9 mmP	P ☆ ○ ☆	Sb, Hg
Finland	9 mmP	SO 44 / ○ / 9	Pb, Sb, Ba
Canada	9 mmP	DI 43 / △ / 9 mm	Pb, Ba
U.S.A.	9 mmP	REM-UMC / ○ / 9 mm LUGER	Pb, Sb, Ba
Sweden	9 mmP	NORMA / ○ / 9 mmP	Pb, Sb, Ba
U.K.	9 mmK	KYNOCH / ○ / ICI	Pb, Sb, Ba
Austria	9 mmK	.380/HP / ★ ○ ★ / 9 mmK ×2	Pb, Ba
Finland	9 mmK	LAPUA / (○) / .380 ACP	Pb, Sb, Ba
Sweden	9 mmK	NORMA / ○ / 380 ACP	Pb, Sb, Ba
Yugoslavia	9 mmK	CAL 9 mm / ○ / PP-74	Pb, Sb
U.S.A.	9 mmK	W-W / ○ / .380 AUTO ×2	Pb, Sb, Ba
Finland	9 mmK	SAKO / ○ / 380 ACP	Pb, Sb
?	9 mmK	FG / ○ / 380 AUTO	Pb, Sb, Ba
Finland	9 mmK	LAPUA / ○ / 9 mm Br Short	Pb, Ba
U.K.	.32 REV	KYNOCH / ○ / 32 S&W	Pb, Sb, Ba
Canada	.380 REV	BROWNING / ○ / 38 S&W	Pb, Sb, Ba
U.K.	.380 REV	KYNOCH / ○ / 38 S+W	Pb, Sb, Ba
U.K.	.380 REV	RG / 380 ○ 66 / 2Z	Pb, Sb, Ba
Austria	.380 REV	HP / ★ ○ ★ / 38 S+W	Pb, Ba
U.S.A.	.380 REV	R ↑L 39 / ○ = / .380	Pb, Sb, Hg
Sweden	.380 REV	NORMA / ○ / 38 S&W	Pb, Sb, Ba
U.S.A.	.380 REV	REM-UMC / ○ / 38 S&W	Pb, Sb, Ba, Hg
Germany	.380 REV	GECO / ⌐ ○ ⌐ / 38 S&W	Pb, Sb, Ba
U.S.A.	.38 SPECIAL	W-W / ○ / 38 SPECIAL	Pb, Sb, Ba
Sweden	.38 SPECIAL	NORMA / ○ / 38 SPECIAL	Pb, Sb, Ba
Germany	.38 SPECIAL	GECO / ○ / 38 SPECIAL	Pb, Sb, Ba
U.S.A.	.357 MAGNUM	R-P / ○ / 357 MAGNUM	Pb, Sb, Ba
U.S.A.	.357 MAGNUM	W-W SUPER / ○ / 357 MAGNUM	Pb, Sb, Ba
U.S.A.	.223 REM	WRA / ○ / 70	Pb, Sb, Ba
U.S.A.	.223 REM	TW / ○ / 66	67 68 +69 72 73 Pb, Sb, Ba
U.S.A.	.223 REM	LC / ○ / 67	69 +70 72 Pb, Sb, Ba
U.S.A.	.223 REM	RA / ○ / 65	66 +67 69 Pb, Sb, Ba
U.S.A.	.223 REM	WCC / ○ / 64	+67 Pb, Sb, Ba
U.S.A.	.223 REM	FC / ○ / 68	Pb, Sb, Ba
U.S.A.	.223 REM	FC / ○ / .223 REM	Pb, Sb, Ba
Canada	.223 REM	IVI / ○ / 70	Pb, Sb, Ba
Sweden	.223 REM	NORMA / ○ / .223	Pb, Sb, Ba
Austria	.223 REM	HP / ★ ○ ★ / 223	Pb, Sb, Ba
U.K.	.380 L	ELEY / ○ / .380 L	Pb, Sb, Hg
Italy	6.5 mm	SMI / ○ / 9-35	Sb

Figure 21.1b (Continued)

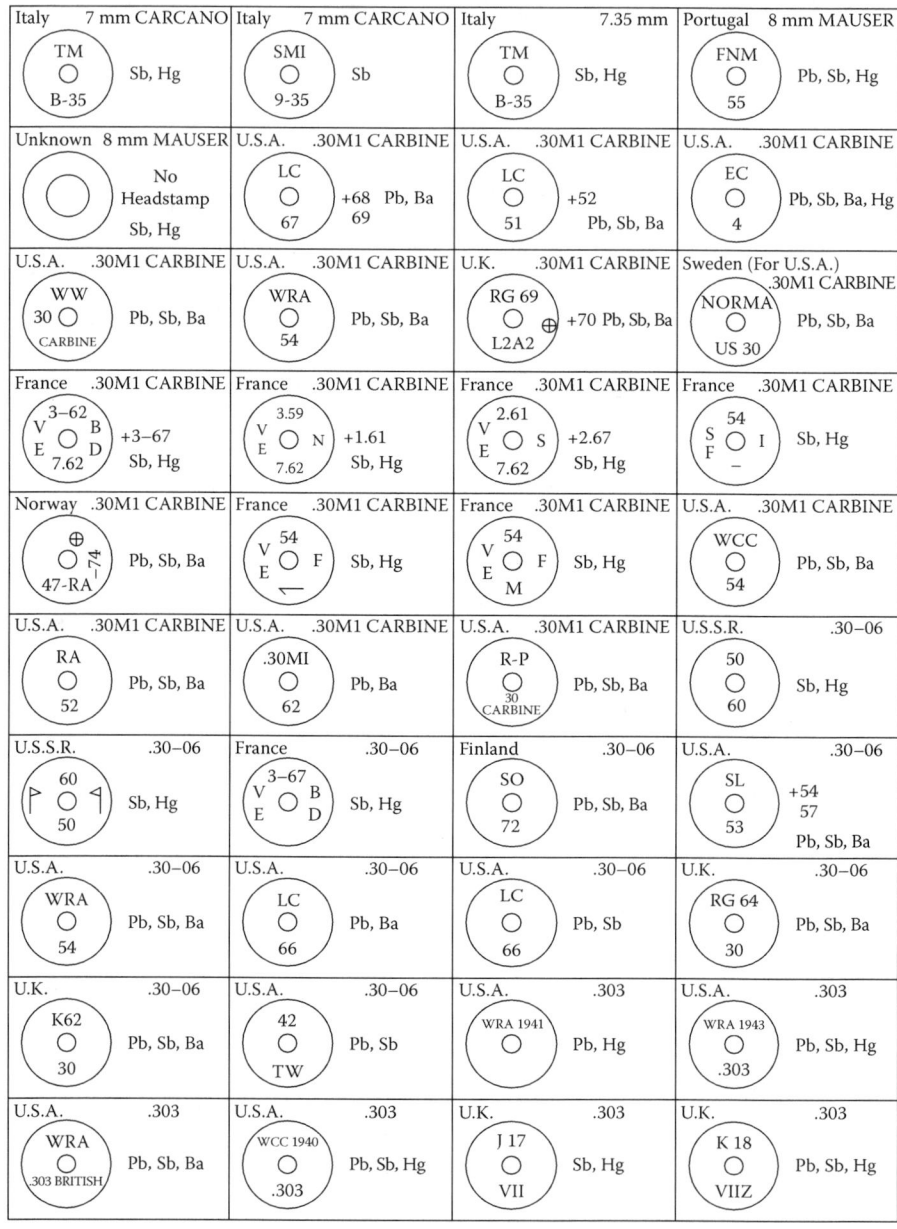

Figure 21.1c (Continued)

Analysis of Ammunition

U.K. .303 K 50 52 ○ +55 Pb, Sb, Hg 7 56 59	**U.K.** .303 (KYNOCH) ○ Pb, Sb, Hg .303	**U.K.** .303 RG ○ Pb, Sb, Hg 54 7	**U.K.** .303 RG 1942 ○ +44 VII Pb, Sb, Hg
U.K. .303 KN 18 ○ Sb, Hg VIIZ	**U.K.** .303 GB 1943 ○ Pb, Sb, Hg VII	**U.K.** .303 G 1918 ○ Sb, Hg IIZ	**U.K.** .303 G-19 ○ Pb, Sb, Ba, Hg VIIZ
U.K. .303 1940 ←○ Pb, Sb, Ba, Hg VII	**Canada** .303 1942 DI ○ N Pb, Ba, Hg VII	**Canada** .303 DC 16 ○ Sb, Hg VII	**Portugal** .303 E ☆○☆ Pb, Sb, Hg 1934
INDIA .303 K↑F VII○ Sb, Hg 11.38	**Czechoslovakia** .303 PSVII 19 ○ 50 Sb, Hg 303	**Belgium** .303 F N ○ Sb, Hg 57	**U.K.** .303 L RG 2 ○ 70 Pb, Sb, Ba A 2 ⊕
U.K. .303 R↑L ○ 49 Pb, Sb, Hg 7	**U.S.A.** .303 WCC 1940 ○ Pb, Sb, Ba, Hg .303	**Syria** 7.62 × 39 mm 9XV17r ★ ○ ★ Sb, Hg 70	**Finland** 7.62 × 39 mm SO ○ Pb, Sb, Ba 72
U.S.S.R. 7.62 × 39 mm 50 ○ Sb, Hg 60	**U.S.S.R.** 7.62 × 39 mm 60 ᑭ ○ ᑫ Sb, Hg 50	**Finland** 7.62 × 39 mm V.P.T. ○ Pb, Sb, Ba 73	**U.S.A.** 7.62 × 51 mm ⊕ ○ 8 6 Pb, Sb, Ba WRA
U.S.A. 7.62 × 51 mm ⊕ ○ Pb, Sb, Ba LC 63	**U.K.** 7.62 × 51 mm 66→70 RG 65 ○ ⊕ +72, 73, 75 L2A2 Pb, Sb, Ba	**U.K.** 7.62 × 51 mm (Tracer) RG 65 ○ Pb, Sb, Ba L5A3	**Norway** 7.62 × 51 mm ⊕ ○ Pb, Sb, Ba 47-RA-74
U.S.A. .308 WIN. MAG. W-W ○ Pb, Sb, Ba SUPER	**France** .45 ACP 56 S ○ I +57 F Sb, Hg 4	**U.S.A.** .45 ACP WCC ○ +69 71 67 Pb, Sb, Ba	**U.S.A.** .45 ACP WRA ○ +65 66 54 Pb, Sb, Ba
U.S.A. .45 ACP WRA-Co ○ Pb, Ba, Hg .45 AC	**U.S.A.** .45 ACP REM-UMC ○ Pb, Sb, Hg .45 ACP	**U.S.A.** .45 ACP RA ○ Pb, Sb, Ba 61	**U.S.A.** .45 ACP REM-UMC ○ Pb, Sb, Ba .45 AUTO
U.S.A. .45 ACP R-P ○ Pb, Sb, Ba 45 AUTO	**France** .45 ACP 57 S ○ I Pb, Sb, Hg F –	**U.S.A.** .45 ACP W-W ○ Pb, Sb, Ba 45 AUTO	**U.K.** .450 REV K II ○ Pb, Sb, Hg C

Figure 21.1d (Continued)

U.K. .450 REV	U.K. .450 REV	U.K. .455 REV	U.K. .455 REV
ELEY-LONDON ○ .450 — Pb, Sb, Hg	ELEY ○ .450 — Pb, Sb, Hg	K ○ .455 2Z — Pb, Sb, Hg	KYNOCH ○ .455 — Pb, Sb, Ba
U.K. .455 REV	U.K. .455 REV	U.K. .455 REV	U.K. .455 REV
K 58 ○ 6Z — 62 +63 64 Pb, Sb, Ba, Hg	K ○ 62 — Pb, Sb, Ba, Hg	K 43 ○ VIZ — Pb, Sb, Hg	R↑L ○ II — 3 6 Pb, Sb, Hg
Canada .455 REV	U.S.A. .50 BROWNING	U.K. 12 BORE SHOTGUN	U.S.S.R. 12 BORE SHOTGUN
DOMINION ○ .455 COLT — Pb, Sb, Ba	RA ○ 41 — Pb, Sb	ELEY-KYNOCH ○ 12 — Pb, Sb, Ba	AZOT ○ 12 12 Made in U.S.S.R. — Pb, Sb, Ba
U.K. 12 BORE SHOTGUN	U.K. 12 BORE SHOTGUN	U.K. 12 BORE SHOTGUN	Italy 12 BORE SHOTGUN
KYNOCH 1 ○ 12 ELEY — Pb, Sb, Ba	ELEY 12 ○ 12 ELEY — Pb, Sb, Ba	GAUGE 12 ○ 12 GAUGE — Pb, Ba	FIOCCHI 12 ○ 12 ITALY — Pb, Sb, Ba
France 12 BORE SHOTGUN	U.S.A. 12 BORE SHOTGUN	Canada 12 BORE SHOTGUN	
GEVELOT 12 ○ 12 PARIS — Pb, Sb, Ba	REMINGTON 12 ○ GA PETERS — Pb, Sb, Ba	C-I-L 12 ○ 12 IMPERIAL — Pb, Ba	

Figure 21.1e (Continued)

Miscellaneous Ammunition Components

Over the past 23 years it has been necessary to examine numerous components of ammunition as a consequence of the requirement of the particular case. Table 21.3 was compiled from casework records and illustrates the variation in, and complexity of, ammunition. It should be noted that "coating" means either a plating or wash and that some cartridge cases and bullet jackets are coated both externally and internally, whereas others are only coated externally. Table 21.3 details the results.

Interpretation of Ammunition Analysis

The quantity of ammunition analyzed provides a good database from which to draw general conclusions. The literature review presented in part 3 is, in the main, supported by Figure 21.1 and Figure 21.2 and Tables 21.1 through Table 21.3.

Brass is the most popular material for cartridge cases and primer cups, with steel the second most widely used material for cartridge case manufacture. Soft copper primer cups encountered were all from old ammunition containing black powder propellant.

Analysis of Ammunition

Country	Cartridge	Headstamp	Elements (primer)	Case composition
U.S.A.	.22 LR	Super	Pb, Ba	(Cu, Si, Na)
U.S.A.	.25 ACP	R-M-U-C / 25 AUTO	Pb, Sb, Ba	(Al, Ca, Cl, Cu, S, Si)
Belgium	.320 REV	FMCEVIL / 3 2 0 EG	Pb, Ba	(Cl, Cu, K, S, Si, Zn)
Germany	.320 REV	SB / 320	Sb, Hg	(Al, Cl, Cu, Fe, K, S, Si, Zn)
Austria	.32 ACP	Hp ★ ★ / 7.65	Pb, Ba	(Ca, Cl, Cu, Fe, K, Si, Zn)
U.S.A.	.32 ACP	R-P / 32 AUTO	Pb, Sb, Ba	(Al, Cu, S, Si)
U.K.	.32 ACP	KYNOCH / 32 AUTO	Pb, Sb, Ba	(Al, Cl, Cu, K, Si, Zn)
Germany	.32 ACP	GECO 7 T / 7.65	Pb, Sb, Ba	(Al, Cl, Cu, S, Si,)
Belgium	.32 ACP	F ★ / N	Pb, Ba	(Cl, Cu, Ni, Zn)
Czechoslovakia	.32 ACP	SBP ☆ ☆ / 7.65	Pb, Sb	(Ca, Cl, Cu, Fe, Zn)
Italy	.32 ACP	GFL / 7.65 mm	Pb, Sb, Ba	(Ca, Cl, Cu, Fe, Ni, Zn)
Canada	.32 ACP	DCC. / 32 ACP	Sb, Hg	(Ca, Cl, Cu, Fe, Zn)
U.S.A.	.32 ACP	E R-M-U-C / 32 / 7.65 mm	Pb, Sb, Ba, Hg	(Ca, Cu, Fe, Zn)
U.S.A.	.32 ACP	W-W / 32 AUTO	Pb, Sb, Ba	(Ca, Cl, Cu, Fe, Ni, Zn)
Sweden	.32 ACP	NORMA / .32 ACP	Pb, Sb, Ba	(Ca, Cl, Cu, Fe)
Finland	9 mmP	VPT / 42	+43 44 Pb, Ba	(Al, Ca, Cl, Cu, Fe, P, S, Si, Zn)
Finland	9 mmP	SO 43 / 9	Pb, Sb, Ba	(Al, Ca, Cl, Cu, Fe, Si, Zn)
Canada	9 mmP	I 4 D 3 / 9 mm	Pb, Ba	(Al, Ca, Cl, Cu, Mg, Si, Zn)
Finland	9 mmP	SO 44 / 9	Pb, Sb, Ba	(Al, Ca, Cl, Cu, K, Si)
Finland	9 mmP	S-41 / 9 mm	Pb, Sb, Ba	(Al, Ca, Cl, Cu, Fe, Si, Zn)
France	9 mmP	78 S I F / 9 mm	Sb, Hg	(Al, Cl, Cu, Fe, K, S, Si, Zn)
Germany	9 mmP	Ch 3 4 T / 91 x	Pb, Sb, Hg	(Al, Cl, Cu, Fe, K, P, S, Si, Sn, Zn)
Germany	9 mmP	GECO / 9 mm	Pb, Sb, Ba	(Al, Ca, Cl, Cu, Fe, K, Mn, S, Si, Sn, Ti, Zn)
Germany	9 mmP	OXO 43 / 9 mm	Pb, Ba	(Al, Ca, Cl, Cu, Fe, K,P, Si)
Germany	9 mmP	17 S / 6	Pb, Sb, Hg	(Cl, Cu, K, Si, Zn)
U.K.	9 mmP	K 58 / 9 mm 2Z	Pb, Sb, Ba	(Al, Ca, Cl, Cu, Fe, K, Si, Zn)
U.K.	9 mmP	RG 59 / 9 mm	+62 Pb, Sb, Hg	(Al, Ca, Cl, Cu, Fe, K, S, Si, Zn)
U.K.	9 mmP	RG 72 / 9 mm	Pb, Sb, Ba	(Al, Ca, Cl, Cu, Fe, K, S, Si, Zn)
U.K.	9 mmP	RG 77 ⊕ / 9 mm 2Z	Pb, Sb, Ba	(Al, Ca, Cl, Cu, Fe, S, Si, Zn)
U.K.	9 mmP	RG 76 / 9 mm 2Z	79 +83 85 Pb, Sb, Ba	(Al, Ca, Cl, Cr, Cu, Fe, K, Si, Zn)
U.S.A.	9 mmP	REM-UMC / 9m/m LUGER	Pb, Sb, Ba	(Al, Ca, Cl, Cu, Fe, Si, Zn)
U.S.A.	9 mmP	WRA / 9 mm	Pb, Sb, Ba	(Ca, Cl, Cu, S, Si, Zn)
U.S.A.	9 mmP	W-W 9mm LUGER	Pb, Sb, Ba	(Al, Ca, Cl, Cu, Fe, K, S, Si, Zn)
U.K.	9 mmP	RG / 9 mm 2Z	+58 Pb, Sb, Ba	(Ca, Cl, Cu, Fe, K, Si, Zn)
Yugoslavia	9 mmP	11 ★ ★ / 52	Pb, Sb, Hg	(Al, Ca, Cl, Cu, Fe, K, Mg, S, Si, Zn)
Germany	9 mmP	9 × 19 G 59 E - / CO 80	Pb, Sb, Ba	(Cl, Cu, Ni)
Sweden	9 mmP	NORMA / 9 mmP	Pb, Sb, Ba	(Ca, Cu, Fe, Ni, Zn)
Czechoslovakia	9 mmP	50 O / ↙ ★	Pb, Sb, Hg	(Al, Ca, Cl, Cu, K, Fe, S, Si, Zn)
Austria	9 mmK	HP ★ ★ / 9 mm	×2 Pb, Ba	(Al, Ca, Cl, Cu, Fe, K, Si, Zn)
Finland	9 mmK	LAPUA / 9 mm SHORT Br	Pb, Sb, Ba	(Al, Ca, Cl, Cu, Fe, Si, Zn)

Figure 21.2a SEM/EDX analysis of spent cartridge cases.

Copper alloy bullet jackets are by far the most common and coated iron jackets are also frequently employed. Lead is by far the most common bullet core material and is often hardened with antimony, but not as often as originally presumed, with antimony occurring in only 25% of the lead bullets examined. Only one of the bullets examined was hardened by tin. Some

Figure 21.2b (Continued)

Country	Caliber	Headstamp	Elements	Trace Elements
Finland	9 mmK	SAKO / 380ACP	Pb, Sb, Ba	(Al, Ca, Cl, Cr, Cu, Fe, S, Si, Zn)
U.K.	9 mmK	ELEY / 380 AUTO	Sb, Hg	(Al, Cl, Cu, Fe, K, S, Si, Zn)
U.S.A.	9 mmK	REM-UMC / 380 CAPH	×2 Sb, Hg	(Al, Cl, Cu, Fe, K, S, Si, Zn)
U.S.A.	9 mmK	WW / 380 AUTO	Pb, Sb, Ba	(Cl, Cu, Fe, S, Si, Zn)
U.S.A.	9 mmK	UMC / 380 CAPH	Sb, Hg	(Cl, Cu, K, S, Si, Zn)
U.S.A.	9 mmK	FC / 380 AUTO	Pb, Sb, Ba	(Al, Cl, Cu, S, Si, Zn)
U.K.	9 mmK	KYNOCH / .380	Pb, Sb, Ba	(Ca, Cl, Cu, Fe, Zn)
Homeload (No Headstamp)	.380 REV		Pb, Ba	(Al, Cl, Cu, Fe, K, Na, S, Si, Ti, Zn)
U.K.	.380 REV	KYNOCH / 38 S+W	Pb, Sb, Ba	??
U.K.	.380 REV	RG 65 / 380 2Z	Pb, Sb, Ba	(Ca, Cu, Fe, K, Si, Ti, Zn)
U.S.A.	.380 REV	RP / .38 S&W	Pb, Sb, Ba	(Al, Cl, Cu, K, Mg, Si, Zn)
U.K.	.380 REV	K 66 / .380 2Z	Pb, Sb, Ba	(Al, Ca, Cl, Cu, Fe, Si, Zn)
U.K.	.380 REV	RG 0 8 6 3 3 2Z	Pb, Sb, Hg	(Al, Cl, Cu, K, P, S, Si, Zn)
U.K.	.380 REV	KYNOCH / 380	Pb, Sb, Ba	(Al, Cu, S)
Sweden	.380 REV	NORMA / 38 S+W	Pb, Sb, Ba	(Ca, Cl, Cu, Fe, Zn)
U.S.A.	.380 REV	REM-UMC / 38 S+W	Pb, Sb, Ba	(Ca, Cl, Cu, Fe, Ni, Zn)
Germany	.380 REV	GECO - r / 38 S+W	Pb, Sb, Ba	(Ca, Cl, Cu, Fe, Ni, Zn)
U.S.A.	.380 REV	BROWNING / 38 S+W	Pb, Sb, Ba	(Ca, Cl, Cu, Fe, Ni, Zn)
Sweden	.38 SPECIAL +P	NORMA / 38 +P SPECIAL	Pb, Sb, Ba	(Al, Ca, Cl, Cu, Fe, Na, S, Si, Zn)
U.S.A.	.38 SPECIAL +P	SPEER W W / 38 SPL +P	Pb, Sb, Ba	(Al, Cl, Cu, Fe, S, Si, Zn)
U.S.A.	.38 SPECIAL	S&W NYCLAD / .38 Spl	Pb, Sb, Ba	(Al, Cl, Cu, Ni, P, S, Si)
U.S.A.	.38 SPECIAL	W-W / .38 SPECIAL	Pb, Sb, Ba	(Cl, Cu, Si, Zn)
Finland	.38 SPECIAL	LAPUA / 38 SPL	Pb, Sb, Ba	(Al, Ca, Cl, Cu, S, Si)
U.S.A.	.38 SPECIAL	R-P / 38 SPL	Pb, Sb, Ba	(Ca, Cl, Cu, Fe, Ni, Zn)
Sweden	.357 MAGNUM	NORMA / .357 MAGNUM	Pb, Sb, Ba	(Al, Cl, Cu, S, Si)
U.S.A.	.357 MAGNUM	SPEER W W / .357 MAGNUM	Pb, Sb, Ba	(Al, Ca, Cl, Cu, Fe, K, Ni, Si, Ti, Zn)
Austria	.223 REM	HP ★ ★ / .223	Pb, Ba	(Al, Ca, Cl, Cu, Fe, Si)
Austria	.223 REM	HP - / 79	Pb, Ba	(Al, Ca, Cl, Cu, Fe, K, Si, Zn)
Canada	.223 REM	IVI / 70	Pb, Sb, Ba	(Al, Ca, Cl, Cu, K, S, Si, Zn)
Sweden	.223 REM	NORMA / .223	Pb, Sb, Ba	(Al, Ca, Cl, Cu, Mg, Si, Zn)
Sweden	.223 REM	NORMA / .223	Sb, Hg	(Al, Cl, Cu, Fe, K, S, Si)
U.S.A.	.223 REM	FC / REM	Pb, Sb, Ba	(Al, Cu, Si, Zn)
U.S.A.	.223 REM	FC / 223 REM	Pb, Sb, Ba	(Al, Ca, Cl, Cu, Fe, S, Si)
U.S.A.	.223 REM	LC / 72 +75 77	Pb, Sb, Ba	(Al, Cl, Cu, Fe, Zn)
U.S.A.	.223 REM	RA / 65 +66	Pb, Sb, Ba	(Al, Ca, Cl, Cu, Fe, S, Si, Zn)
U.S.A.	.223 REM	TW / 67	Pb, Sb, Ba	(Al, Ca, Cl, Cu, Fe, K, Si, Zn)
U.S.A.	.223 REM	TW / 72 +73	Pb, Sb, Ba	(Al, Cu, Fe, K, Mn, S, Si, Zn)
U.S.A.	.223 REM	WCC / 64	Pb, Sb, Ba	(Al, Cl, Cu, Fe, Si, Zn)
U.S.A.	.223 REM	WRA / 70	Pb, Sb, Ba	(Al, Cl, Cu, Fe, K, S, Si)
Italy	6.5 mm CARCANO	SMI / 935	Pb, Sb, Hg	(Ca, Cl, Cu, Fe, K, S, Si, Zn)

combination of the elements barium, strontium, magnesium, iron, and chlorine were present in tracer bullets.

The review suggests that diphenylamine (DPA) is the most common stabilizer in single-based propellants, whereas ethylcentralite (EC) is the most common in double based. In fact, Table 21.1 and Table 21.2 show that DPA

Analysis of Ammunition

Country	Caliber	Headstamp	Primer	Propellant additives
France	.30 M1 CARBINE	54 / S F / I / -	Sb, Hg	(Al, Cl, Cu, Fe, K, S, Si, Sn, Zn)
France	.30 M1 CARBINE	54 / VE / F / r	Pb, Sb, Ba	(Ca, Cl, Cu, Fe, K, Si)
France	.30 M1 CARBINE	54 / VE / F / 1	Sb, Hg	(Al, Cl, Cu, Fe, K, Ni, S, Si, Zn)
France	.30 M1 CARBINE	56 / VE / F / n	Sb, Hg	(Al, Cl, Cu, Fe, K, S, Si, Zn)
France	.30 M1 CARBINE	3.59 / VE / N / 7.62	Sb	(Al, Cl, Cu, Fe, K, S, Si)
France	.30 M1 CARBINE	1.61 / VE / N / 7.62	Sb, Hg	(Cl, Cu, Fe, K, S, Si, Zn)
France	.30 M1 CARBINE	2.61 / VE / S / 7.62	Pb, Sb, Ba	(Al, Ca, Cl, Cu, Fe, K, Si, Zn)
France	.30 M1 CARBINE	2-61 / VE / S / 7.62	Sb, Hg	(Al, Cl, Cr, Cu, Fe, K, S, Si, Zn)
France	.30 M1 CARBINE	3.61 / VE / S / 7.62	Sb	(Al, Cl, Cu, K, S, Si)
France	.30 M1 CARBINE	3-62 / VE / B D / 7.62	Sb, Hg	(Al, Cl, Cu, Fe, K, S, Si, Zn)
France	.30 M1 CARBINE	4.63 / VE / S / 7.62	Pb, Sb, Hg	(Al, Cl, Cu, Fe, K, S, Si, Zn)
France	.30 M1 CARBINE	1-63 / VE / S / 7.62	Pb, Sb, Hg	(Al, Ca, Cl, Cu, Fe, K, S, Si, Zn)
France	.30 M1 CARBINE	3.67 / VE / B D / 7.62	Sb, Ba, Hg	(Al, Ca, Cl, Cu, Fe, K, S, Si, Zn)
Sweden	.30 M1 CARBINE	NORMA / US.30	Pb, Sb, Ba	(Al, Ca, Cl, Cu, Fe, K, S, Si, Zn)
U.S.A.	.30 M1 CARBINE	WCC / 42	Pb, Sb, Ba	(Al, Cl, Cu, Fe, K, S, Si, Zn)
U.S.A.	.30 M1 CARBINE	W-W / .30 / CARBINE	Pb, Sb, Ba	(Al, Cl, Cu, Fe, S, Si, Zn)
France	.30 M1 CARBINE	3.67 / VE / B D / 7.62	Sb, Hg	(Ca, Cl, Cu, Fe, K, Si, Zn)
France	.30 M1 CARBINE	54 / VE / F / 5	Pb, Sb, Ba	(Ca, Cl, Cu, Fe, K, Si)
France	.30-06	4.53 / T / S / 7.62	Sb, Ba, Hg	(Al, Cl, Cu, Fe, K, S, Si, Zn)
France	.30-06	1.55 / T H / S / 7.62	Pb, Sb, Ba, Hg	(Al, Ca, Cl, Cu, Fe, K, S, Si, Zn)
France	.30-06	4-54 / T H / S / 7.62	Pb, Sb, Ba, Hg	(Al, Ca, Cl, Cu, Fe, K, Mg, S, Si, Zn)
Italy	.30-06	P / B D / 953	Pb, Sb, Hg	(Cl, Cu, Fe, K, S, Si, Zn)
U.K.	.30-06	K 60 / 7	Pb, Sb, Hg	(Al, Cl, Cu, Fe, K, S, Si, Zn)
U.S.A.	.30-06	E / D / N / 44	Pb, Sb	(Al, Cl, Cu, K, S, Si, Zn)
U.S.A.	.30-06	FA / 27	Pb, Sb	(Al, Cl, Cu, Fe, K, S, Si, Zn)
U.S.A.	.30-06	LC / 43	Pb, Sb	(Al, Cl, Cu, Fe, K, Si, Zn)
U.S.A.	.30-06	LC / 53	Sb	(Cl, Cu, K, S, Zn)
U.S.A.	.30-06	RA / 55	Pb, Sb, Ba	(Al, Ca, Cl, Cu, Fe, Si, Zn)
U.S.A.	.30-06	SL / 42	Pb, Sb, Hg	(Al, Cl, Cu, Fe, K, S, Si, Zn)
U.S.A.	.30-06	SL / 53	Pb, Sb, Ba	(Al, Cl, Cu, Fe, K, Si, Zn)
Belgium	.303	FN / 57	Sb, Hg	(Al, Cl, Cu, Fe, K, S, Si, Sn, Zn)
Canada	.303	DC 16 / VM	Sb, Hg	(Al, Cl, Cu, K, S, Si, Zn)
Canada	.303	1943 / DIZ	Pb, Ba	(Al, Ca, Cl, Cu, Fe, Si, Zn)
Italy	.303	P / B D / 953	Pb, Sb, Hg	(Al, Cl, Cu, Fe, K, S, Si, Zn)
U.K.	.303	1941 / ← / VII	Pb, Sb, Hg	(Cl, Cu, K, S)
U.S.A.	.303	WRA 1940 / .303	+41 Pb, Sb	(Al, Cl, Cu, K, Na, S, Zn)
Portugal	.303	FNM / 50	Sb, Hg	(Cl, Cu, Fe, K, S, Si, Sn, Zn)
U.K.	.303	K 60 / 7	Pb, Sb	(Al, Cl, Cu, Fe, K, S, Si)
Czechoslovakia	.303	PS VII / 19 50 / .303	Sb, Hg	(Al, Cl, Cu, Fe, K, S, Si, Sn, Zn)
China	7.62 × 39 mm	31 / 69	+70 Sb, Hg	(Al, Cl, Cu, Fe, K, P, S, Si, Sn, Zn)

Figure 21.2c (Continued)

and/or its derivatives occur in the majority of propellants: ~94.5% of single based and ~82.5% of double based. Ethyl centralite and DPA frequently occur together, whereas EC on its own is only found in ~2.0% of single based and ~14.5% of double based. It must be remembered that these are "discharged" propellants, their origin is largely unknown, and the figures are

Czechoslovakia 7.62 × 39 mm b×h / 54 / Pb, Sb, Hg (Al, Cl, Cu, Fe, K, Si, Zn)	Czechoslovakia 7.62 × 39 mm B× / 53 h ×2 / 1 / Pb, Sb (Al, Cl, Cu, Fe, K, Mn, S, Si, Zn)	Finland 7.62 × 39 mm SO / 72 / Pb, Sb, Ba (Al, Ca, Cl, Cu, Fe, K, S, Si, Zn)	Finland 7.62 × 39 mm VPT / 69 / Pb, Sb, Ba (Al, Ca, Cl, Cu, Fe, P, Si, Sn, Ti, Zn)
Middle East 7.62 × 39 mm ∠ ▲ / 7.62 × 39 / Pb, Sb (Al, Cl, Cu, Fe, K, Mn, P, S, Si, Zn)	Syria 7.62 × 39 mm PTXVL ★ ○ ★ F / 7 B / Sb, Hg (Al, Cl, Cu, Fe, K, S, Sn, Zn)	Syria 7.62 × 39 mm x9x t, T ★ ○ ★ / 70 / Sb, Hg (Cl, Cu, Fe, K, S, Si, Sn, Zn)	U.S.S.R. 7.62 × 39 mm 60 ▷ ○ ◁ / 50 / Sb, Hg (Al, Cl, Cu, Fe, K, S, Si, Sn, Zn)
U.S.S.R. 7.62 × 39 mm 5.39 ★ ○ ★ / 58 / Sb (Al, Cl, Cu, Fe, K, Mn, S, Si, Sn, Zn)	U.S.S.R. 7.62 × 39 mm 60 ★ ○ ★ / E / Sb, Hg (Al, Cl, Cu, Fe, K, S, Si, Sn, Zn)	France 7.62 × 51 mm 4-54 T ○ S E / 7.62 / Sb, Hg (Al, Ca, Cl, Cr, Cu, Fe, K, S, Si, Ti, Zn)	France 7.62 × 51 mm 2.60 V ○ S E / 7.62 / Sb, Hg (Al, Cl, Cu, K, S, Si, Zn)
U.K. 7.62 × 51 mm RG 71 ○ ⊕ +76 / L5A5 / 77 / Pb, Sb, Ba (Al, Ca, Cl, Cu, Fe, K, Si, Zn)	U.K. 7.62 × 51 mm RG 77 ⊕ ○ / 7.62 / Pb, Sb, Ba (Al, Cl, Cu, Fe, K, Mg, Si, Zn)	Sweden 7.62 × 51 mm ⊕ 48 ○ 74 ×2 -RA- / Pb, Ba (Al, Ca, Cl, Cu, Fe, K, Si, Sn, Zn)	Yugoslavia 7.92 × 57 mm 11 ★ ○ ★ / 52 / Pb, Sb, Ba, Hg (Al, Ca, Cu, Fe, P, S, Si, Sn, Zn)
France .45 ACP 56 S ○ I F / 4 / +57 Sb, Hg (Al, Cl, Cu, Fe, K, S, Zn)	U.S.A. .45 ACP REM-UMC ○ / .45 ACP / Pb, Sb, Ba (Al, Ca, Cl, Cu, S, Si, Zn)	U.S.A. .45 ACP REM-UMC ○ / .45 ACP / Pb, Sb (Al, Ca, Cl, Cu, Fe, K, S, Si, Sn, Zn)	U.S.A. .45 ACP RA ○ / 42 / Pb, Sb, Ba, Hg (Al, Ca, Cl, Cu, Fe, K, S, Si, Sn, Zn)
U.S.A. .45 ACP RP ○ / 45 AUTO / Pb, Sb, Ba (Al, Ca, Cl, Cu, Fe, Na, Ni, S, Si, Zn)	U.S.A. .45 ACP RA W ○ Co / .45 A.C. / Pb, Ba, Hg (Al, Ca, Cl, Cu, S, Si, Zn)	U.S.A. .45 ACP C W ○ C / 71 / Pb, Sb, Ba (Al, Ca, Cl, Cu, Fe, K, S, Si, Zn)	U.S.A. .45 ACP T W ○ / 5 / Pb, Ba (Cl, Cu, Fe, Ni)
U.S.A. .45 ACP C E ○ S / 43 / Pb, Sb (Cl, Cu, K, Fe, Ni, Zn)	U.S.A. .45 ACP W-W ○ / 45 AUTO / Pb, Sb, Ba (Cl, Cu, K)	U.S.A. .45 ACP RA ○ / 62 / Pb, Ba (Al, Cl, Cu, Ni, Si, Zn)	U.S.A. .45 ACP WRA ○ / 66 / Pb, Sb, Ba (Cl, Cu, Fe, Ni, Zn)
U.K. .450 REV E Y L O L ○ N E O D 450 N / Pb, Hg (Al, Cl, Cu, Si, Zn)	U.K. .455 REV K C ○ II / Pb, Sb, Hg (Cl, Cu, Fe, K, S, Si, Zn)	U.K. 12 BORE SHOTGUN E Y K Y L ○ O E O C 12 H / Pb, Sb, Ba (Al, Ca, Cl, Cu, Fe, S, Si, Zn)	U.K. 12 BORE SHOTGUN SPECIAL 12 ○ 12 SMOKELESS / Pb, Sb, Ba (Ca, Cl, Cu, Fe, K, S, Si)
U.K. 12 BORE SHOTGUN ELEY INTERNATIONAL GAME 6 2.6 mm 12 ★ ○ ★ 12 / Pb, Sb, Ba (Al, Ca, Cl, Cu, K, Fe, Mn, Ni, S, Si, Zn)	U.K. BATON ROUND 1.5"/38 mm 12/77 ○ FPL BLACK POWDER 46 CY5.79 L5A3 / Pb, Ba (Al, Cu, K, S, Si)	U.K. BATON ROUND 1.5"/38 mm 7/73 ○ FPL BLACK POWDER 25L5A3 SPRA 7-75 / Pb, Ba (Al, Cu, K, S)	U.K. BATON ROUND 1.5"/38 mm 2/81 ○ FPL BLACK POWDER RUBBER BATON Mk 2 / Pb, Ba (Al, Ca, Cl, Cu, K, S, Si)

Figure 21.2d (Continued)

local to Northern Ireland; consequently the percentages must be viewed with caution. Methyl centralite occurs in ~8% of the propellants and according to the literature it may be used either as a plasticizer or moderant. It always occurred accompanied by other plasticizers and is more likely to be included as a moderant rather than as a plasticizer. Tributylphosphate was detected in three propellants and it is not mentioned in the literature. Its function is uncertain but it is used as a plasticizer in certain industrial processes. On the other hand, many of the compounds mentioned in the review as possible

Analysis of Ammunition

Table 21.1 Analysis of Single-Based Propellants (NC present)

No. of Shooting Incidents	DPA	NDPA	EC	MC	DBP	DNT	Comments
40	√	—	—	—	—	—	
10	√	—	√	—	—	—	
5	√	—	—	—	—	—	+ Camphor
4	√	√	√	—	—	—	
4	√	—	—	√	—	—	
4	√	—	—	—	√	√	
3	√	—	—	—	—	√	+ Naphthalene in one of the samples
2	√	—	√	—	—	√	
2	—	—	—	—	—	—	Only camphor detected
2	√	—	—	—	—	√	+ Camphor (benzene and naphthalene also detected in one sample)
2	√	√	—	—	—	—	+ Sulfur
2	√	—	—	√	—	√	
2	√	×2	—	—	√	—	Tributylphosphate also detected in one sample
1	—	×2	√	—	√	√	
1	—	—	—	—	√	√	+ Cresol
1	√	√	—	—	—	√	+ Camphor and naphthalene
1	√	√	√	—	√	—	
1	√	√	√	—	—	—	+ Camphor
1	—	—	√	√	—	—	
1	√	√	—	—	—	—	+ MEDPA
1	—	×2	√	—	√	√	+ Naphthalene
1	√	×2	√	—	√	—	
1	—	—	√	—	—	—	

Key: DBP, dibutylphthalate; DNT, dinitrotoluene; DPA, diphenylamine; EC, ethylcentralite; MC, methylcentralite; MEDPA, methylethyldiphenylamine; NC, nitrocellulose; NDPA, a nitrodiphenylamine; NG, nitroglycerine; TNT, trinitrotoluene; √, detected; —, not detected.

constituents of propellants were not detected in the propellants analyzed. However, it must be borne in mind that Table 21.1 and Table 21.2 represent only a small selection of propellants, which have been discharged and which are local to Northern Ireland.

Table 21.2 Analysis of Double-Based Propellants (NC, NG present)

No. of Shooting Incidents	DPA	NDPA	EC	MC	DBP	DNT	Comments
20	√	—	—	—	—	—	Two NDPAs also detected in one sample
12	—	—	√	—	—	—	Naphthalene also detected in one sample
12	√	—	√	—	—	—	
11	√	—	—	—	—	√	Sulfur, TNT, Tributylphosphate also detected, occurring separately in three samples
10	√	—	√	—	—	√	
5	√	—	—	—	√	√	
5	√	—	—	—	√	—	
4	√	—	√	√	—	√	Naphthalene also detected in one sample
3	√	—	√	—	√	√	Two NDPAs also detected in one sample
2	√	√	—	—	√	√	
2	√	√	—	—	—	—	
2	√	√	—	—	—	√	
2	√	—	√	√	—	—	
1	—	√	√	—	—	—	
1	√	√	√	√	√	√	
1	√	√	√	—	—	—	Tributylphosphate also detected
1	√	—	√	—	√	—	
1	—	—	—	—	—	—	Only camphor detected
1	—	—	—	—	√	—	
1	—	—	—	—	√	√	
1	—	√	—	—	√	—	
1	—	—	√	√	√	—	
1	—	—	√	—	√	—	
1	—	—	√	—	—	√	
1	√	—	—	√	—	√	

See Table 21.1 for abbreviations.

Analysis of Ammunition

Table 21.3 Analysis of Ammunition Components

Description	Observations
.22, Eley, standard velocity, primer .22 tracer bullet	Pb, Ba, (Al, Ca, Cu, P, Si) Pb core with Ba, Sr, Mg containing tracer composition
German military bullet (no detail)	Brass-coated Fe jacket/Fe core
12 bore, metal base of case	Many are Sn-coated Fe, e.g., Eley
Silvalube bullet (Mountain & Sawden)	Al-coated Pb (trace Sb, Cu, Fe, Si)
9 mmP, MEN-83-25, primer	Pb, Sb, Ba (Al, Ca, Cl, Cu, Fe, K, Si, Sn, Zn)
9 mmP, OXO43, bullet	Brass-coated Fe jacket/Fe core
9 mmP, K582Z, bullet	Brass-coated Fe jacket/Pb core
9 mmP, Ch (Belgian), bullet	Brass-coated Fe jacket/Pb sheath, Fe core
9 mmP, SF178, bullet	Brass jacket/Pb core (trace Sb)
9 mmP, 1X*51.2, bullet	Ni-coated Fe jacket/Fe core with Pb sheath
9 mmP, Mauser, bullet	Ni (trace Cu)-coated Fe (trace Mn) jacket/Pb core (trace Sb)
.30 M1, Norma triclad, bullet	Cu/Ni-coated Fe jacket/Pb core (trace Sb)
.38 S&W, Kynoch, bullet	Unjacketed Pb (trace Sb)
.38 Equaloy, bullet	Al bullet; outside skin (Al, Ti, major: Cl, Fe, P, S, trace), inside Al only
.38SPL, Winchester Silvertip, bullet	Al jacket/Pb core
.38SPL, W-W Lubaloy, bullet	Cu jacket/Pb core
.38SPL, W-W, bullet	Unjacketed Pb (trace Sb)
.38SPL, Kynoch, bullet	Unjacketed Pb (trace Sn)
.357 Mag, KTW metal piercing bullet	Homogeneous brass with green plastic (Teflon) coating containing Al, Cr, Ti, major: Ca, Cl, Cu, K, S, Si, Zn trace
.357 Mag, W-W Super, bullet	Solid brass with exposed top portion coated with green plastic (Teflon) containing Cr, Ti
.450, Eley, bullet	Unjacketed Pb
.450, Kynoch, bullet	Unjacketed Pb (trace Sb)
.455, Kynoch, bullet	Unjacketed Pb
7.62 NATO, RAUFOSS, bullet	Brass jacket/Pb core (trace Sb)
7.62 NATO, 47-RA-77, tracer bullet	Cu jacket/Sr and Fe at tail end: Pb sheath, Fe core at nose end
.308 WIN, PMC, primer	Pb, Sb, Ba (Al, Ca, Cl, Cu, K, S, Si)
7.9 mm, Mauser, bullet	Cu/Ni-coated Fe jacket/Pb core (trace Sb)
.30-06, SL53, AP bullet	Brass jacket/Pb sheath, Fe core (trace Mn)
12 bore, Eley International, primer	Pb, Sb, Ba (Al, Ca, Cl, Cu, K, Fe, Mn, Ni, S, Si, Zn) Note: Fe frequently at major level

Table 21.3 Analysis of Ammunition Components (Continued)

Description	Observations
12.7 mm, Russian 188/83 AP/I bullet	Cu jacket (trace)/Pb sheath, Fe core: incendiary powder contained Mg, Al, Ba
7.62 NATO, L5A3, tracer bullet	Tracer composition contained Cl, Cu, Sr (trace Al, Ba, Bi, Ca, Fe, K, Ni, S, Si, Zn)
.50 tracer bullet	Tracer composition contained Ba with a trace of S
223, 84.SF, SFM, round	Frangible bullet with Cu/Sn, Al case, Pb, Sb, Ba primer, single base propellant (DPA)
.223 NATO, RORG88, ROTA round	Frangible bullet with Cu/Si/W, brass case, Pb, Sb, Ba primer, double-based DPA, DBP
.25 AUTO, REM-UMC, round	Cu jacket/Pb core, brass case, Ni-coated brass primer cup
.25 AUTO, R-P, round	Cu jacket/Pb core, double-based propellant
.32 AUTO, GECO LT, round	Ni-coated brass jacket/Pb core, brass case
.32 AUTO, RWS, round	Unjacketed Pb bullet, brass case, Cu primer cup, black powder
.32-20, UMC, round	Lubricated unjacketed Pb bullet, brass case, Cu primer cup, black powder
.297-,230, no head stamp, round	Unjacketed Pb bullet, brass case, Cu primer cup, black powder with fiber wad
9 Mk, *HP*, round	Fe jacket/Pb core (trace Sb, Fe, Cu, Al, Si), Al-coated steel case, Ni-coated brass primer cup; propellant contains K and S; Pb, Sb, Ba primer
9 mmP, CCI.NR, round	Cu jacket/Pb core (trace Sb, Fe, Cu, Al, Si), Al-coated steel case, Ni-coated brass primer cup; propellant contains K and S; Pb, Sb, Ba primer
9 mmP, ELEY 83, round	Ni-coated brass primer cup, brass case (trace Al), Pb, Sb, Ba primer
9 mmP, SBP, round	Fe jacket/Pb core (trace Si, Fe); brass primer cup, anvil and case; Pb, Ba primer (trace Sn)
9 mmP, B↑E43, round	Cu jacket/Pb core, brass case
9 mmP, *11,50,9, round	Ni-coated Fe jacket (trace Zn)/Fe core, brass primer cup and case; Pb, Sb, Hg primer
9 mmP, NATO, RG85 round	Brass jacket/Pb core (trace Si), brass primer cup and case
9 mmP, W-W, round	Ni-coated brass jacket/Pb core (trace Si), brass case, Ni-coated brass primer cup
9 mmP, 12* 49×51, round	Fe jacket/Fe core with Pb sheath, brass case
9 mmP, SFM-THV, round	Solid brass bullet, brass case; Pb, Ba primer
9 mmP, SANDIA, round	Brass jacket/Pb core (trace Si), brass case; Pb, Sb, Ba primer

Analysis of Ammunition

Table 21.3 Analysis of Ammunition Components (Continued)

Description	Observations
9 mmP, R.P, round	Cu jacket/Pb core, brass case, Ni-coated brass primer cup
9 mmP, GECO*, round	Brass-coated Fe jacket/Pb core, brass case, Ni-coated brass primer cup
9 mmP NATO, RG84.2Z, round	Cu jacket/Pb core, brass case and primer cup
9 mmP NATO, FFV88, round	Cu-coated Fe jacket/Pb core, brass case, and primer cup
9 mmP NATO, FNM84-12, round	Cu-coated Fe jacket/Pb core, brass case, and primer cup
9 mmP, NORMA round	Cu-coated Fe jacket/Pb core, brass case, Ni-coated brass primer cup
9 mmP, WIN, round	Cu jacket/Pb core, brass case, Ni-coated brass primer cup
9 mmP, S&B, round	Ni-coated Fe jacket/Pb core, brass case, Ni-coated brass primer cup; Pb, Sb, Ba primer (trace Sn)
.380 AUTO, W-W, round	Brass jacket/Pb core (trace Si); brass case, Ni-coated brass primer cup; Pb, Sb, Ba primer (trace Sn)
.380 REV, R↑L345.2Z, round	Cu-coated Fe jacket/Pb core, brass case, and primer cup
.380 REV, K66.2Z, round	Cu jacket/Pb core, brass case, and primer cup
.38 S&W, Kynoch, round	Unjacketed Pb bullet (trace Sb), brass case, and primer cup
.38 S&W, Kynoch, round	Lubricated unjacketed Pb bullet, brass case, Ni-coated brass primer cup
.38 SPL, SBW, round	Unjacketed Pb bullet (trace Al, Ca, Si); brass case, primer cup, and anvil; Pb, Sb, Ba primer
.38 SPL, S&W, round	Pb bullet fully coated with plastic (Teflon), Ni-coated brass case
.38 SPL, LAPUA, round	Lubricated unjacketed Pb bullet, brass case and primer cup
.38 SPL, W-W, round	Cu jacket/Pb core (trace Sb), Ni-coated brass case and primer cup
.38 SPL, NORMA, round	Cu jacket/Pb core, brass case, Ni-coated brass primer cup
.38 SPL, R.P, round	Lubricated unjacketed Pb bullet, Ni-coated brass case
.38 SPL, CCI.NR, round	Lubricated unjacketed Pb bullet (trace Sb), Al case (trace Cu, Fe, Mn)
.38 SPL+P, W SUPER W, round	Al jacket/Pb core, Ni-coated brass case

Table 21.3 Analysis of Ammunition Components (Continued)

Description	Observations
.38 SPL+P, W-SUPER-W, round	Lubricated unjacketed Pb bullet, Ni-coated brass case and primer cup, brass anvil
.38 SPL +P, CCI.NR, round	Cu jacket/Pb core (trace Sb), Al case, brass primer cup (trace Fe); Pb, Sb, Ba primer
.38 SPL +P, SFM-THV, round	Solid brass bullet, brass case; Pb, Sb, Ba primer
.357 MAG, CCI.NR, round	Al case, Ni-coated brass primer cup; Pb, Sb, Ba primer
.357 MAG, NORMA, round	Brass jacket/Pb core; brass case, primer cup, and anvil; Pb Sb, Ba primer
.357 MAG, W-W SUPER, round	Unjacketed Pb bullet (trace Si); brass case, primer cup, and anvil; Pb, Sb, Ba primer
.357 MAG, W-W SUPER, round	Brass jacket/Pb core (trace Sb), Ni-coated brass case (trace Al)
.357 MAG, W-W SUPER, round	Cu jacket/Pb core, Ni-coated brass case and primer cup
.45 ACP, WRA 68, round	Cu-coated Fe jacket/Pb core, brass case and primer cup
.45 ACP, WCC73, round	Cu-coated Fe jacket/Pb core, brass case and primer cup
.45ACP, SF14.56, round	Brass jacket/Pb core, brass case and primer cup
.45 ACP, FN45*, round	Cu jacket/Pb core, brass case and primer cup
.45ACP, RA68, round	Cu-coated Fe jacket/Pb core, Ni-coated brass case and primer cup
.45ACP, R.P., round	Cu jacket/Pb core, Ni-coated brass case
.45ACP, W-W, round	Cu jacket/Pb core, brass case
.30 Mauser, Kynoch, round	Cu jacket/Pb core ("K" marked on base), brass case
.30MI, DAG.VL, round	Cu-coated Fe jacket/Pb core, brass case
7.9 STEYER, no head stamp, round	Ni-coated Fe jacket/Pb core, brass case
.223, FN79, AP round	Cu jacket/Steel penetrator with Pb sheath, brass case
.223, FNB83, round	Cu jacket/Pb core with steel tip, brass case, and primer cup
.223, TW72, round	Cu jacket/Pb core, brass case, and primer cup
.223, LC72, round	Cu jacket/Pb core, brass case, and primer cup
7.62 × 39, VPT73 round	Cu jacket/Pb core, brass case
7.62 × 39, BXN51, round	Cu-coated Fe jacket/steel core with Pb sheath, lacquered steel case, brass primer cup
7.62 NATO, RG84, round	Cu jacket/Pb core, brass case

Table 21.3 Analysis of Ammunition Components (Continued)

Description	Observations
7.62 NATO, 47-RA-74, round	Cu jacket/Pb core, brass case
7.62 NATO, FN78, round	Cu jacket/steel penetrator with Pb at base, brass case
7.62 × 51, FLB78, round	Brass jacket (trace Ni)/Pb core, brass (trace Al) case and primer cup; Pb, Sb, Ba primer
7.62 × 51, 89-070, AP round	Cu-coated Fe jacket/hardened steel core with base enclosed in Al cup; brass case and primer cup
.30-06, K53 round	Cu-coated Fe jacket/Pb core, brass case and primer cup
.30-06, K58, round	Cu-coated Fe jacket/Pb core, brass case, and primer cup
.30-06, DM42, round	Cu-coated Fe jacket/Pb core, brass case, and primer cup
.30-06, FA54, round	Cu-coated Fe jacket/Pb core, brass case, and primer cup
7.62 NATO, 12-RA-78, tracer bullet	Cu-coated Fe jacket/Pb nose, tracer composition contains Sr, Mg, Cl, tracer igniter composition contains Sr, Cu with minor Zn, base enclosed with Cu disc
.38 SPL, WCC, primer	Pb, Sb, Ba (Al, Ca, Cu, Fe, K, Mn, Ni, S, Si, Ti, Zn)
.357 MAG, FEDERAL, primer	Pb, Sb, Ba (Ca, Cu, K, Fe, Mn, Ni, S, Si, Ti, Zn)
.357 MAG, HP, primer	Pb, Sb, Ba (Ca, Cu, K, Fe, Ni, S, Si, Ti, Zn)
9 mmP NATO, RG83, primer	Pb, Sb, Ba (Al, Ca, Cu, Fe, S, Si, Ti, Zn)
.32ACP, S&B primer	Pb, Sb, Ba (Ca, Cu, Fe, K, Mn, Ni, P, S, Si, Ti, Zn)
7.65 mm, GECO, primer	Pb, Sb, Ba (Al, Ca, Cu, Fe, K, Mn, Ni, S, Si, Sn, Ti, Zn)
.30 MI, DAG, primer	Pb, Sb, Ba (Ca, Cu, Fe, K, Mn, P, S, Si, Ti, Zn)
.30MI, WINCHESTER, primer	Pb, Sb, Ba (Al, Ca, Cu, Fe, K, Mn, S, Si, Ti, Zn)
TW 72, .223" caliber ammunition	Cu, Zn jacket/Pb core (trace Al); Cu, Zn primer cup; Pb, Sb, Ba primer; double-based propellant with DPA, DNT, a phthalate plasticizer
WRA70, .223" caliber ammunition	Cu, Zn jacket/Pb core (trace Sb); Cu, Zn primer cup; Pb Sb, Ba primer; double-based propellant with DPA, DNT, a phthalate plasticizer
LC 72, .223" caliber ammunition	Cu (trace Al) jacket/Pb core (trace Al); Cu, Zn primer cup; Pb, Sb, Ba primer; double-based propellant with DPA, DNT, a phthalate plasticizer

Table 21.3 Analysis of Ammunition Components (Continued)

Description	Observations
FNB 83, .223" caliber ammunition	Cu, Zn jacket/Pb core (trace Al), Fe (trace Al) penetrator; Cu, Zn primer cup; Pb, Sb, Ba primer; double-based propellant with DPA, DNT, a phthalate plasticizer

Note: See Glossary for firearms/ammunition-related abbreviations.

The majority of the propellants analyzed were from kneecapping incidents involving the use of handguns. A total of 194 propellant samples were analyzed of which 92 were single based. Rimfire cartridges and rifles are rarely used in kneecappings.

Ammunition recovered in Northern Ireland covers a time span of more than 50 years of ammunition manufacture, and many residues remaining in the spent cartridge case have been analyzed. The residue examined does not necessarily originate exclusively from the primer, as a contribution could be made by the propellant or the exposed base of the bullet. However, the residue appears to reflect the primer type in the majority of instances. A more satisfactory way to determine primer type is to remove and open the spent primer cup and examine the inside using SEM/EDX. This is not practical in casework as it would mean the destruction of evidence and would be time-consuming, tedious, and in the vast majority of instances, unnecessary. A more satisfactory method is to remove the bullet and propellant from a live round, discharge the primer, and then sample the spent cartridge case interior. However, in casework the actual spent cartridge cases involved in the incident are sampled, as it cannot be assumed that ammunition of the same caliber and head stamp will have the same composition.

Information obtained from visits to various munitions factories suggests that manufacturers will use whatever is available at the time, from whatever source, to complete an order, provided that it meets the required ballistics performance and produces no residues that are injurious to the gun. During the war years, shortage of material meant many variations in materials used in manufacture. For these reasons it is unwise to make assumptions about ammunition components and composition, even for the same caliber and manufacturer, as they could vary from batch to batch. The differences between ammunition with the same head stamp can be seen in Table 21.3 for Winchester Western in .38 Special and .357 Magnum calibers and in Figure 21.2 for .30 M1 caliber VE 54 F1 and VE 2-61 S.

The analysis of "primers" supports the statement that Communist Bloc countries frequently use mercury fulminate primers and it is also worth noting that the same applies to ammunition manufactured in France, at least for the time period involved. According to the literature there has been no mercury in U.S. military ammunition since 1898, but it was used to a later date (~1930) in some U.S. commercial primers. Although the ammunition data in Figure 21.1 and Figure 21.2 strongly support this, there are some anomalies, namely, 30-06 caliber SL-42, .30M1 caliber EC4, .303" caliber WRA 41 and 43 and WCC 1940, and .45 ACP caliber RA 42, all of which had mercury present and were manufactured during war years.

A detailed summary of primer types encountered over a 13-year period (1975 to 1987) is given in Table 21.4 which represents the examination of 1,300 spent cartridge cases, involving 310 different head stamps and 58 manufacturers, and is based on casework results, some of which are included in Figure 21.1 and Figure 21.2 and Table 21.3.

Primers could be grouped into six categories: (a) corrosive and mercuric (potassium chlorate and mercury fulminate), (b) noncorrosive and mercuric (barium nitrate replaced potassium chlorate), (c) corrosive and nonmercuric (lead styphnate replaced mercury fulminate), (d) noncorrosive, nonmercuric (modern Sinoxyd type), (e) unusual/miscellaneous primer compositions, and (f) recent nontoxic primers (Sintox).

The fact that the spent cartridge cases in Table 21.4 were not analyzed for potassium or chlorine to indicate potassium chlorate makes interpretation difficult. Nevertheless, Table 21.4 does support the history of primer development as outlined in Chapter 9.

Category (f) Sintox primers can be excluded from consideration as they were introduced at a later date and their use has not yet been encountered in casework. For category (a) mercury would be present and barium would be absent. From Table 21.4 approximately 76.5% of mercury-containing primers are corrosive. For category (b) both mercury and barium would be present. Therefore, approximately 23.5% of mercury-containing primers are noncorrosive.

For category (c) mercury and barium would be absent and lead would be present. Modern type primers would be lead, antimony, barium, and lead, barium. On this basis a somewhat speculative breakdown of primer types involved in casework during this period is presented:

~67.5% modern
~24.0% mercury fulminate
~6.0% nonmercuric but corrosive
~2.5% miscellaneous

This also supports the history of primer development.

These figures reflect the situation prior to 1988. As a consequence of terrorist organizations on both sides acquiring large arms consignments, since March 1988 the IRA and related groups frequently use the 7.62 × 39 mm caliber AKM type rifle with Yugoslavian nny 82 ammunition whereas the

Table 21.4 Analysis of Spent Cartridge Cases for Pb, Sb, Ba, Hg

Caliber	Pb/Sb/Ba	Pb/Ba	Pb/Sb	Sb/Hg	Pb/Ba/Hg	Pb/Ba/Sb/Hg	Pb/Sb/Hg	Pb	Sb	Pb/Hg	Ba/Sb/Hg	No. of Head Stamps
.22 LR	2	37	4	—	—	—	—	5	—	—	—	7
.32 ACP	42	7	3	—	—	—	—	—	—	—	—	1
9 mmP	103	15	3	26	—	7	31	—	—	—	3	58
9 mmK	46	34	7	8	—	—	—	—	—	—	—	28
.380 Rev	18	4	—	—	—	2	2	—	—	—	—	10
.38 Special	12	—	—	—	—	—	—	—	—	—	—	5
.223	278	3	11	1	—	—	—	3	—	—	—	25
8 mm Mauser	—	—	—	1	—	—	8	—	—	—	—	2
.30 MI	34	12	1	34	—	32	10	—	27	—	2	31
.30-06	49	2	18	4	—	2	5	—	1	—	2	21
.303	4	1	3	15	1	6	28	—	—	2	—	34
7.62 × 39 mm	16	—	9	33	—	—	—	4	3	—	—	14
7.62 × 51 mm	54	3	1	2	—	—	—	—	—	—	—	20
.45 ACP	61	3	1	1	2	1	2	—	—	—	—	16
.450 Rev	—	—	—	—	—	—	9	—	—	3	—	3
.455 Rev	3	1	—	—	—	11	9	—	—	—	—	11
12 Bore	24	7	3	—	—	—	—	—	—	—	—	10
Miscellaneous	57.35	—	2	—	—	1	—	2	1	1	—	5
Total	749	129	66	125	3	62	104	14	35	6	7	310
~%	57.5	10.0	5.0	9.5	<0.5	5.0	8.0	1.0	3.0	0.5	0.5	

Note: See Glossary for firearms/ammunition-related abbreviations.

loyalist paramilitary groups use the 7.62 × 39 mm caliber VZ 58P rifle with Chinese 351/73 ammunition.

Analysis of the nny 82 ammunition as previously detailed shows that it uses a mercuric corrosive primer. Analysis of the Chinese 351/73 ammunition revealed that it has a copper-coated iron-jacketed bullet with an iron core and a lead tip, the cartridge case is steel with a brown colored lacquered finish, a brass primer cup, and the propellant is single based with DPA, 2 × nitrodiphenylamines, camphor, and contains no inorganic additives, and the discharged primer composition is antimony, potassium, chlorine, mercury, tin, sulfur, iron, manganese, phosphorus, zinc, and lead in descending order (lead, antimony, mercury type).

The frequent use of both types of ammunition in recent years has substantially increased the proportion of shooting incidents involving the use of mercury fulminate–primed ammunition. It has also substantially increased the proportion of shooting incidents involving the use of single-based propellant. It is worth noting that both propellants contain camphor.

Reference

199. T. G. Kee, D. M. Holmes, K. Doolan, J. A. Hamill, and R. M. E. Griffin, "The Identification of Individual Propellant Particles," *Journal of the Forensic Science Society* 30 (1990): 285.

Ammunition Containing Mercury

22

The particle classification scheme, developed as described in reference 200, did not include mercury fulminate–primed ammunition, which is frequently encountered in Northern Ireland, and is currently manufactured in some Eastern Bloc countries.

In casework in which discharge residue particles were detected, and in which the ammunition involved is known to contain mercury, very few, if any, of the particles contained mercury. This has been noted over many years and in numerous cases. Possible reasons for this could be the volatility of mercury and its compounds, or decomposition of the mercury fulminate and the loss of mercury through amalgamation with zinc in the primer cup/cartridge case. It is not uncommon, when firing old ammunition with mercury fulminate primers, for some of the cartridge cases to crack, due to embrittlement of the brass caused by mercury amalgamating with the zinc.

To clarify the situation regarding mercury-containing ammunition a series of experiments were conducted and casework statistics gathered.

Frequency of Occurrence

To determine the frequency of occurrence of mercury-containing particles in FDR, promptly collected residue from the discharge of mercury fulminate–primed ammunition was examined. Results are given in Table 22.1. In the first firing a small proportion of the particles also contained one of the following elements: cobalt (trace), magnesium (trace), nickel (trace), and phosphorus (minor and trace).

The particle types containing mercury that were detected in firings 1 to 7 (Table 22.1) are as listed in Table 22.2. As can be seen from Table 22.1, even with promptly collected residue, the proportion of mercury-containing particles is very low. It is interesting to note that firing 1, the ammunition with the unjacketed bullet, produced fewer lead-only particles than expected but did produce a large proportion of lead, antimony particles. This suggests the majority of the lead, antimony particles originated from the bullet rather than from the primer.

On other occasions mercury-containing ammunition has been test fired, the hands sampled immediately, the samples analyzed by SEM/EDX, and

Table 22.1 Occurrence of Mercury Particles

Ammunition	Pb, Sb, Ba	Sb, Ba	Ba, Ca, Si	Pb, Sb	Pb, Ba	Pb Only	Sb Only	Ba Only	Brass	Fe Major	Others	Hg	Total No. of Particles
1. Unjacketed K.455 2Z Pb, Sb, Ba, Hg	1 <0.5%	None —	None —	218 71.0%	2 0.5%	45 14.5%	1 <0.5%	1 <0.5%	2 0.5%	17 5.5%	3 1.0%	16 5.0%	306
2. FMJ (Cu) K64 6Z Pb, Sb, Ba	22 11.0%	1 0.5%	None —	30 15.0%	1 0.5%	88 45.0%	1 0.5%	3 1.5%	6 3.0%	33 17.0%	11 5.5%	None —	196
3. FMJ (Cu) K58 6Z Pb, Sb, Ba, Hg	7 2.0%	None —	None —	193 65.0%	4 1.0%	36 12.0%	3 1.0%	3 1.0%	8 2.5%	33 11.0%	None —	8 2.5%	295
4. FMJ (Cu) RG56 9 mm 2Z Pb, Sb, Ba, Hg	8 2.5%	None —	2 0.5%	13 4.5%	8 2.5%	139 47.0%	None —	1 <0.5%	9 3.0%	99 33.5%	15 5.0%	1 <0.5%	295
5. FMJ (Cu) RG59 9 mm 2Z Pb, Sb, Ba, Hg	11 3.5%	1 <0.5%	1 <0.5%	24 8.0%	20 6.5%	103 34.5%	2 0.5%	None —	25 8.0%	81 27.0%	27 9.0%	3 1.0%	298
6. FMJ (Cu) RG55 9 mm 2Z Pb, Sb, Ba, Hg	14 5.0%	None —	3 1.0%	42 14.0%	9 3.0%	101 34.5%	None —	1 <0.5%	30 10.0%	63 21.5%	23 8.0%	5 1.5%	291
7. FMJ (Cu) K56 9 mm 2Z Pb, Sb, Ba, Hg	24 8.0%	2 0.5%	10 3.5%	39 13.0%	27 9.0%	84 28.5%	1 <0.5%	1 <0.5%	35 12.0%	52 18.0%	19 6.5%	1 <0.5%	295

Note: See Glossary for firearms/ammunition-related abbreviations.

Ammunition Containing Mercury

Table 22.2   Mercury-Containing Particles

Major	Minor	Trace	Particle Type
Pb	Si, Sb	Cl, K, Cu, Fe, Hg	Pb, Sb, Hg
Pb, Sb	—	Cu, K, Si, Cl, Zn, Hg, Fe	Pb, Sb, Hg
3 × Pb	Si	Cl, Sb, Cu, Hg	Pb, Sb, Hg
Pb	Sb, K, Cl, Si, Al	Cu, Zn, Hg	Pb, Sb, Hg
Pb	Sb, Si, Al	Cu, Hg	Pb, Sb, Hg
Pb	Sb, Cl, K, Si	Cu, Fe, Zn, Al, Hg	Pb, Sb, Hg
Pb, S, Sb, K, Cl	Si, P, Al, Cu	Hg, Fe	Pb, Sb, Hg
Sb, Pb, S	Cu, Si, Al, K, Cl	Fe, Zn, Hg	Pb, Sb, Hg
Si, Al	K, Ca, Fe	Hg	Hg only
Pb, Cl	Sb, K, Si	Cu, Co, Hg	Pb, Sb, Hg
Pb, S	Sb, Si	Cu, Fe, Hg	Pb, Sb, Hg
Pb, S, Cl, K, Sb	Si	Cu, Hg	Pb, Sb, Hg
Pb, S	Si, Sb, Cl	K, Mg, Cu, Fe, Hg	Pb, Sb, Hg
2 × Pb, S, Sb	Cu	Cl, Hg	Pb, Sb, Hg
Pb, Cl, Sb	K	Cu, Si, Fe, Hg	Pb, Sb, Hg
Pb, S, Sb	Cl, K	Cu, Fe, Hg	Pb, Sb, Hg
Pb, S	Sb, Cu	Si, K, Hg, Fe, Cl	Pb, Sb, Hg
Pb, S, Sb	Cu	Si, Hg, Fe, Cl	Pb, Sb, Hg
2 × Pb	Sb	Cu, Cl, K, Fe, Hg	Pb, Sb, Hg
Pb	Sb, Hg	Cu, Si	Pb, Sb, Hg
2 × Pb, Sn	Cu, Zn	Hg, Si	Pb, Hg (Sn)
Ba, Ca, Si	Pb, S	Cl, K, Cu, Hg	Pb, Ba, Hg
Pb, S, Sb, Ba	Si, K, Cl, Cu	Zn, Fe, Hg	Pb, Sb, Ba, Hg
Pb, S, Sb, Ca, Cu	Ba, Cl, K, Si, Fe	Zn, Hg	Pb, Sb, Ba, Hg
Pb	Si, Sb, Hg	Cl, Cu	Pb, Sb, Hg
Pb, S	Sb, Hg, Cu	Si, Zn, Fe	Pb, Sb, Hg
Pb, Sb	Hg, Cl	Si, Fe, Cu, Zn	Pb, Sb, Hg
Pb, S, Sb	Hg	Cu, Si, Cl	Pb, Sb, Hg
Pb, S, Sb	K, Cl	Cu, Hg	Pb, Sb, Hg

mercury was not detected in any of the particles. One such test involved the firing of ammunition which, according to the sampling of the interior of the spent cartridge case, had antimony and mercury in the primer. No mercury was detected in any of the discharge particles but lead and barium were detected in some particles. The lead is thought to originate from the base of the bullet (FMJ) and the barium from inorganic additives to the propellant.

This supports the proposition that anything present in a round of ammunition can make a contribution to the composition of the discharge residue particles. The presence of tin in any of the discharge particles is an indication that the primer contains mercury, the tin originating from the tinfoil disc used to seal mercury fulminate. (Tin is also present in some modern ammunition components, for example, Sellier & Bellot, and it is present in some propellants.)

Mercury-Containing Particles in Casework

Particles containing mercury are relatively rare in the environment; the only noncartridge source previously detected was from dental fillings. A brief search of the literature on the uses of mercury revealed that it is used for fungicides/bactericides, amalgamation, catalysts, special solders (along with lead and tin), dental preparations, electrical apparatus, electrolytic preparation of chlorine and sodium hydroxide, government/commercial laboratory use, paint (antifouling, mildew proofing), paper pulp (slime inhibitor), pharmaceuticals (ointments, antiseptics, diuretics), manufacture of thermometers and barometers, timber preservation, photography, and in the manufacture of mercury fulminate. A representative selection of mercury-containing particles detected in casework involving the use of mercury fulminate–primed ammunition is presented in Table 22.3 and is included to demonstrate the variety of compositional types.

Distribution of Mercury after Discharge

A specific ammunition type was examined to determine the total mercury content. Results are given in Table 22.4.

Spent cartridge cases, originating from the same ammunition type, were examined to determine the amount of mercury remaining in the cartridge case. Results are given in Table 22.5.

Using the same ammunition type, the amount of mercury deposited on the perimeter of a bullet hole was then determined. Results are given in Table 22.6.

The amount of mercury remaining on the bullet after it had passed though a target was then determined. Results are presented in Table 22.7.

The amount of mercury remaining in the gun after discharge was then determined. The results are given in Table 22.8.

These tests give an estimate of the total amount of mercury in a complete round of ammunition prior to discharge (4,070 µg), the amount deposited on the bullet (2.15 + 16.78 = 18.93 µg), the amount remaining in the spent

Ammunition Containing Mercury

Table 22.3 Casework Particles Containing Mercury

Major	Minor	Trace	Particle Type
Si, Pb, Fe	Ca, K, Cl	Cu, Hg	Pb, Hg
Sn	Cu	Zn, Hg	Hg only
Sb, Sn, Cl, S	—	Fe, Cu, Hg	Sb, Hg (Sn)
Hg	Cu, Zn	Cl, K, Ca	Hg only
Hg, Cu, Zn	Sb, Sn	Cl	Sb, Hg
Hg	—	Cu, Zn	Hg only
Hg	—	Cu, Zn, Si, Cl	Hg only
Hg	Cl, K	Sb, Cu	Sb, Hg
Hg, S, Sb	K, Cl, Cu	Si	Sb, Hg
Hg	Cl, K, Cu, Sb	Zn, Si	Sb, Hg
Sb	Cu, Hg, Cl	—	Sb, Hg
Sb, Hg	Cl, Cu	K, Zn	Sb, Hg
Hg	Sb, Cu, Cl, K	Zn, Si	Sb, Hg
Sb, Cu, Hg	Cl	K	Sb, Hg
Sb, Cl, Cu, S, Hg	K, Si	Zn, Al	Sb, Hg
K, Cl	Hg, S, Cu	Sb, Si	Sb, Hg
Si, Pb	Ca, Cl	Al, P, Hg, K, Ti	Pb, Hg
Hg	Ca, Si	Al, K, Fe, Cu, Zn, Ti, Mg	Hg only
K	Hg, S, Sb, Cu	Si, Zn	Sb, Hg
Sb, Cl	K, Hg, S, Cu	Si, Al, Zn	Sb, Hg
Si	Ca, Hg, Ag, Al	Fe, Cu	Dental filling
Pb, Sb, Hg, Ca	Si, Cu, Cl, Fe, Al	K, Mg, Cr, Ti	Pb, Sb, Hg
Hg, Sb	Cu	K, Cl, Zn	Sb, Hg
Sb	Hg, Cu	Cl, Si	Sb, Hg
Cu, Cl	Hg, K, Sb, Zn	—	Sb, Hg
Cu, Hg, Sb	K, Cl	Zn	Sb, Hg
Cu, S, Sb	K, Cl	Zn, Al, Hg	Sb, Hg
Hg, Cu	Zn, K, Cl	Si, Fe	Hg only
Hg, Cl	Cu, Zn	K, Si	Hg only
Cu, Zn	Cl, Sb, Pb, Hg	K, Si	Pb, Sb, Hg
Cu	—	Cl, Hg, Pb, Zn	Pb, Hg
Hg, Pb, Cu, Sb, Cl	—	Zn	Pb, Sb, Hg
Hg, S, Cl, K, Sb	Cu	—	Sb, Hg
Hg	—	Cl, K, Sb	Sb, Hg
Cl, K, Hg	Sb	Cu	Sb, Hg

Table 22.3 Casework Particles Containing Mercury (Continued)

Major	Minor	Trace	Particle Type
Sb, Hg, S	Cu	Fe, Cl	Sb, Hg
Sb, Hg, S	K	Ba, Fe	Sb, Ba, Hg
Sb	K, Hg, S, Cl	—	Sb, Hg
Sb, Sn	Cu	Hg, Cl	Sb, Hg (Sn)
Sb	S, Sn	Cu, Hg, Al	Sb, Hg
Cl, K	Hg, S, Sb, Cu	Si, Zn	Sb, Hg
Cl, K, S, Hg, Sb	Cu, Si	Al, Zn	Sb, Hg
Sb, K, Cl, Pb, S	Cu, Al, Hg	—	Pb, Sb, Hg
Pb, S, K, Sb, Cl	Al, Cu	Si, P, Hg, Zn	Pb, Sb, Hg
Cu, Cl, Zn, Hg, Pb, S	K, Sb	—	Pb, Sb, Hg
Pb, S	Sb, Cl, K, Si, Hg	Cu, Zn, Fe	Pb, Sb, Hg
Pb, Sb	Hg, Cu	Si, Al	Pb, Sb, Hg
Hg, Si, Ca	Fe, Cu, Zn, K, Al, Mg, Cl	—	Hg only
Sb, Si	S, Cl, K, Ti, Al, P, Fe	Hg, Cu, Zn	Sb, Hg
S, Pb, Hg, Sb, Si	Fe, Al, K	Cu, Cl, Mg	Pb, Sb, Hg
Hg, S	Al, Si, K, Ca, Fe, Cu	—	Hg only
Pb, Hg, Sb, Ca	Si, Cu, Cl, Fe, Al	K, Mg, Cr, Ti	Pb, Sb, Hg
Pb, Hg, Cu, Sb	K, Cl, Ba	Al, Zn	Pb, Sb, Ba, Hg
Cu, Sb	Pb, Hg	Cl, K, Al	Pb, Sb, Hg
Sb, Cl, Cu, S, Hg	K, Si	Zn, Al	Sb, Hg
Hg	Si, Ca	Al, Fe, Cr, K, Cu, Zn	Hg only
Hg	Cl, K, Cu, Sb	Zn, Si	Sb, Hg
Hg, S	Si	Ca, Ti, Mg, Fe, Cu	Hg only
Pb, Sb	Ca, Hg, Cl, Cu	K, Fe	Pb, Sb, Hg
Hg, Ca	Si, Al	K, Fe, Cu, Zn	Hg only

cartridge case (533 µg), and the amount remaining in the gun (6.5 µg). This gives a difference of ~3,511 µg (85%) of mercury released into the environment during the discharge process. To determine what proportion of this would be detectable by SEM/EDX a further set of experiments was devised. The initial test involved the examination of residue from the discharge of a primer, in a primed cartridge case (no propellant or bullet present), to determine what proportion of the mercury would be detectable in 1 µm or greater, particulate form (Figure 22.1). Results are given in Table 22.9.

Ammunition Containing Mercury

Table 22.4 Total Mercury Content

Sample No.	Mercury µg	Average µg
1	3,850	
2	3,500	
3	3,900	
4	4,150	
5	4,200	
6	4,550	
7	4,450	
8	4,150	4,070
9	4,050	
10	3,750	
11	4,000	
12	4,450	
13	4,500	
14	3,950	
15	3,600	

Table 22.5 Mercury Remaining in Spent Cartridge Case

Sample No.	Mercury µg	Average µg
1	490	
2	710	
3	555	
4	470	
5	465	
6	695	
7	550	
8	405	533
9	680	
10	520	
11	510	
12	545	
13	460	
14	510	
15	425	

Table 22.6 Mercury in Bullet Wipe

Sample No.	Mercury μg	Average μg
1	1.32	
2	2.95	
3	3.00	
4	1.85	
5	1.51	2.15
6	2.10	
7	2.23	
8	2.17	
9	2.35	
10	1.97	

Table 22.7 Mercury Remaining on Bullet

Sample No.	Mercury μg	Average μg
1	11.6	
2	20.8	
3	10.2	
4	20.8	16.78
5	13.9	
6	17.7	
7	16.3	
8	17.7	
9	14.2	
10	14.6	

Table 22.8 Mercury Remaining in Gun

Sample No.	Mercury μg	Average μg
1	5.7	
2	7.1	6.5
3	6.8	
4	6.3	

Ammunition Containing Mercury

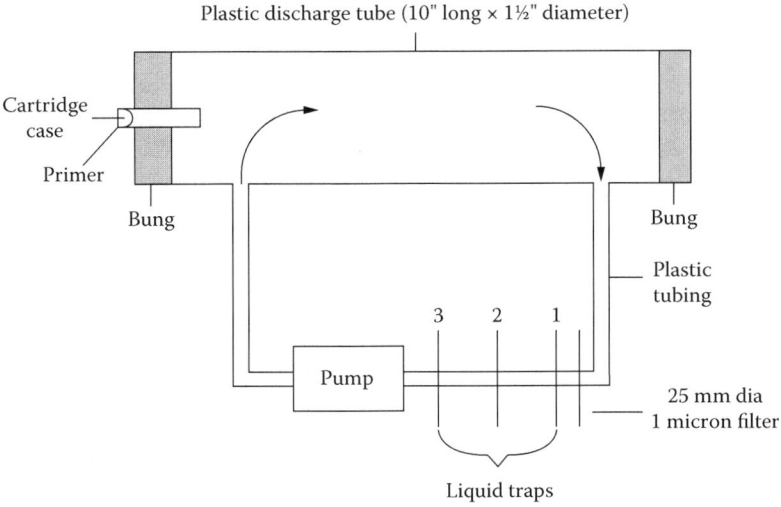

Figure 22.1 Primer discharge sampling system.

Table 22.9 Mercury Levels from Primer Discharge

			Liquid		
Test No.	Cartridge Case (μg)	Filter (μg)	Trap 1 (μg)	Trap 2 (μg)	Trap 3 (μg)
1	969.0	51.6	225.0	23.6	8.1
2	1,030.0	50.9	271.0	30.3	9.7
3	941.0	73.0	277.0	28.2	7.9

Average results indicate that approximately 24% of the mercury remains in the spent cartridge case/primer cup, approximately 7% of the mercury was recovered from the liquid traps, and only 1.5% of the mercury was present on the filter. Approximately 68% of the mercury appears to be present as large particulate matter, which must have been deposited on the interior of the firing tube.

The mercury concentration remaining in the spent cartridge case was considerably higher than previously experienced. This test did not reproduce the conditions experienced during the discharge of a round of ammunition, where much higher temperatures and pressures are attained, plus the possible suction effect in the wake of the bullet. To simulate actual conditions a further test was devised (Figure 22.2).

This test involved the discharge of a complete round of ammunition and facilitated the examination of discharge residue exiting the muzzle of the firearm. The primary objective of the examination was to determine what

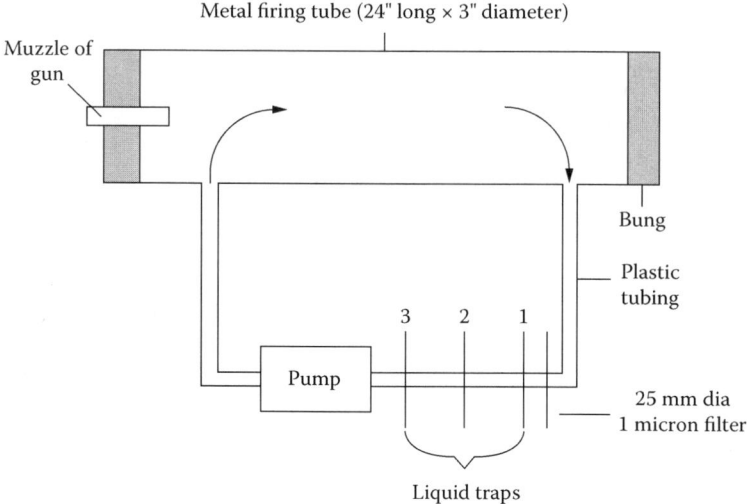

Figure 22.2 Muzzle discharge sampling system.

proportion of the mercury-containing discharge residue particulate matter is likely to be detected by SEM/EDX. Results are given in Table 22.10.

The proportion of the discharge residue issuing from the muzzle that has the potential to be detectable by SEM is 12.1%, 11.7%, and 13.5%, respectively. These figures are remarkably reproducible given the scope for experimental error in this type of experiment. The distribution between the filter and the liquid traps, that is, between particulate and vapor, is 17.7%, 17.2%, and 20.2%, respectively, on the filter.

Finally, breech discharge residue was examined to investigate the occurrence of mercury-containing particles that would be SEM/EDX detectable (Figure 22.3) using the same pump, filter, and liquid traps as the previous experiment. Results are given in Table 22.11.

Averaging the results it would appear that in the region of 6.60 µg of mercury exits via the breech (ignoring spent cartridge case), of which approximately 48% is particulate (approximately 40% was retained on the filter).

Table 22.10 Mercury Levels in Muzzle Discharge Residue

Test No.	Inside Tube (µg)	Filter (µg)	Liquid		
			Trap 1 (µg)	Trap 2 (µg)	Trap 3 (µg)
1	444.0	171.0	786.0	6.92	3.66
2	521.0	194.0	919.0	10.70	4.10
3	403.0	165.0	641.0	6.12	4.11

Ammunition Containing Mercury

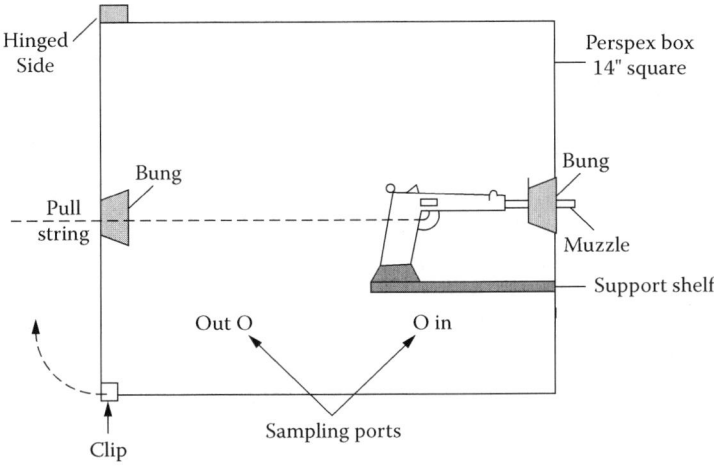

Figure 22.3 Sampling box for breech residue.

Table 22.11 Mercury in Breech Discharge Residue

			Liquid		
Test No.	Inside Box (µg)	Filter (µg)	Trap 1 (µg)	Trap 2 (µg)	Trap 3 (µg)
1	0.68	2.05	2.56	0.51	0.11
2	0.50	3.01	2.97	0.55	0.11
3	0.57	2.73	2.55	0.73	0.19

Table 22.12 gives a quantitative summary of the measured distribution of mercury after discharge. Percentages are based on an initial amount of 4,070 µg.

A percentage recovery of 48.73 is disappointing. The variation in the amount of mercury initially present in the ammunition (see Table 22.4) is probably due to deterioration of the mercury fulminate over a long period, the ammunition having been manufactured in 1943. The variation would

Table 22.12 Distribution of Mercury After Discharge

Distribution	Mercury (µg)	~%
Spent cartridge	533.00	13.10
Bullet wipe	2.15	0.05
Spent bullet	16.78	0.40
Residue in gun	6.50	0.16
Exits muzzle	1,412.00	34.86
Exits breech	6.60	0.16
Totals	1,977.03	48.73

also be a major factor in accounting for the varying amounts remaining in the spent cartridge case. Ideally a recently manufactured batch lot of ammunition should have been used but none was available, and to acquire some in the United Kingdom in a pistol caliber would be very difficult. From the start it was known that the variation in mercury levels in the ammunition would introduce large errors in the percentage recovery, somewhere in the region of ±16%. However, the objective was to explain the scarcity of mercury-containing particles from discharge residue, as the primary concern was the recovery of particulate matter using a 1-μm pore size membrane filter.

The data confirm that, as expected, the vast majority of residue exits from the muzzle, only 12.4% of which was detected on the membrane filter and 55.5% in the liquid traps, the balance probably remaining in the apparatus. The most important aspect is the discharge residue exiting from the breech, as it is some of this residue that is likely to be deposited on the firer. Only 6.6 μg of mercury was detected (excluding spent cartridge case) of which 39.1% was detected on the membrane filter and 52.0% in the liquid traps, the balance presumably remaining in the firing box. The presence of the spent cartridge case in the firing box may have caused, or contributed to, the amount of mercury remaining in the box.

The primer only discharge test produced interesting results: only approximately 1.5% of the mercury was retained by the 1-μm filter. The spent cartridge cases had a significantly higher level of mercury than when they were discharged normally in a firearm. This is thought to be due to the very substantial temperature and pressure difference between the discharge of non-bulleted and bulleted ammunition, coupled with the absence of a possible suction effect in the wake of a bullet. The physical disturbance of extraction and ejection of the spent cartridge case and the shock it suffers on striking the ground could dislodge mercury-containing particles from its surface. In the test only the inside surface of the cartridge case was exposed to discharge residue due to the fact that the outer surface was surrounded by a rubber bung, whereas in normal use in a firearm the outside surface would also be exposed to a residue-laden environment.

The low concentration of mercury on the filter is surprising given that only the initial source, the primer, was involved plus the smaller volume of the firing tube, compared to the other experiments, which should have increased sampling efficiency. This tends to suggest that the majority of the mercury was contained in large particles that were deposited on the inside of the firing tube, because the particles are too large to be circulated by the pump.

The primer is the origin of the mercury; the average amount is 4,070 μg, which is equivalent to approximately 5,775 μg mercury fulminate. This is burned at high temperature and pressure, with the vast majority of it expelled with force from the muzzle and away from the firer. The pistol used in the tests works on the blow back principle. This involves the use of gas pressure

Table 22.13 Quantitative Comparison between Filter and Liquid Traps

	Filter		Liquid Traps	
	Amount Hg (µg)	%	Amount Hg (µg)	%
Primer discharge	58.50	17	293.60	83
Muzzle residue	176.70	18	793.90	82
Breech residue	2.60	43	3.43	57

to move the slide to the rear, thereby extracting the spent cartridge case from the chamber. The ejector then expels the spent cartridge case through the ejection port and the slide moves forward again under spring pressure and, in doing so, chambers another round from the magazine. It is the blow back "gases" that are the most likely to be deposited on the firer under normal circumstances. Table 22.13 gives a comparison between quantities detected on the filter and liquid traps.

There is remarkable agreement between the primer discharge and muzzle residue, filter to liquid ratios. The breech residue shows a marked difference which is difficult to explain, the filter retaining a much higher percentage of the mercury. It is possible that the filter picked up some larger particles, with high mercury content, from the ejection of the spent cartridge case. This is unlikely and would need to occur in all three tests.

There was debris inside the firing tubes after discharge of the primers and bullets. It was a sooty deposit which contained particulate matter. All the filters were dirty after use, with the residue gray/black in color, and no large particles were noted on the surface of the filter on macroscopic visual examination. There are three distinct forms of residue involved: large particles that were not transported through the sampling system, smaller particles that were transported through the sampling system and that were greater than 1 µm in size, and vapor/submicron particles that passed through the filter and were retained in the liquid traps.

It is obvious that there is some form of concentration effect of the particulate matter in the breech emission. When a round of ammunition is discharged in a firearm there is considerable pressure generated, which acts in all direction including rearward. It is the rearward pressure of the cartridge case acting on the breech face which drives the slide backward. This backward pressure may also tend to contain the bulk of the primer particles, which are at the start of the discharge process, and create a zone of discharge residue, traveling rearward, which is rich in particles originating from the primer. Whatever the reason the residue exiting from the breech has a higher concentration of small detectable particles than the muzzle exhaust.

From the tests it is clear that a large amount of mercury is present in vapor/submicron particle form and is consequently not detectable by SEM.

A large proportion of the mercury is also present in large particulate matter, which is very unlikely to be deposited on the firer as it exits from the muzzle and travels a considerable distance away from the firer. Only 2.6 μg mercury that exited from the breech and possibly 0.55 μg present in the sampling box (which may have originated from the spent cartridge case) are potentially detectable on the firer. In practice only a small proportion of this will be deposited on the firer as it will disperse in all directions. Given the low percentage of mercury-containing particles detected in promptly collected residue from indoor firings under favorable laboratory conditions, it is not surprising that very few mercury-containing particles are detected in casework.

The nature of the tests performed is such that large experimental errors are involved. Care was taken to ensure that all the apparatus was airtight and all manipulations, extractions, washings, and so forth were done as carefully as possible. Despite the care taken, a low percentage recovery of mercury resulted. Some mercury was probably lost by adsorption/absorption onto/into the metal (firing tube), rubber (bungs and gaskets), plastic (tubing), and glass (liquid traps). Some may have been lost through the pistol after discharge, as the pistol would not be airtight. The rinsing of the firing tube, after discharge and suction sampling, was difficult and probably very inefficient and no rinsing of the glass tubing in the liquid traps or plastic connecting tubing was attempted. The possibility of particles present in the wake of the bullet exiting the firing tube along with the bullet cannot be excluded. It is likely that a small amount of mercury would have been lost from the surface of the bullet into the bullet recovery medium.

The pump used was not very powerful (18 liters/minute). This was by choice, as a more powerful pump could possibly have caused larger particles to enter the sampling system, which would not reflect a real-life situation, as the larger particles would not remain airborne for long.

Despite the experimental errors it is obvious that a high percentage (86%) of the mercury is released into the atmosphere, the majority via the muzzle. The majority (88%) that exits the muzzle is not detectable by SEM. A very small percentage exits via the breech (0.16%) of which only approximately 40% is detectable by SEM.

Indications are that the majority of the mercury is present in submicron particulate/vapor form and in large particulate form which is unlikely to be deposited on the firer. All these observations account for the scarcity of discharge residue particles containing mercury. The scarcity of mercury-containing discharge particles on the firer has been confirmed by other workers.[201]

A surprising result was the amount of mercury remaining on the bullet after it had passed through a wad of filter paper and the bullet recovery medium. Mercury was also readily detectable in the bullet hole perimeter tests (Table 20.12 and Table 20.13).

Ammunition Containing Mercury

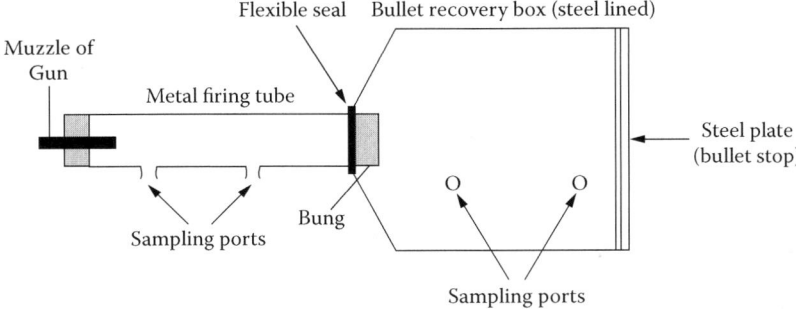

Figure 22.4 Improved residue sampling apparatus.

The possibility of relating bullet damage to a particular round of ammunition by chemical comparison of the deposit on the bullet hole perimeter with the deposit on the spent bullet should be investigated.

It would be desirable to repeat the experiments using recently manufactured batch lots of ammunition and testing for lead, antimony, and barium in addition to mercury. It would also be desirable to examine a range of ammunition and handguns, rifles being difficult, although not impossible, to accommodate because of size and pressure generated on discharge. Unfortunately time did not permit. An improved design of apparatus would incorporate the choice of a closed or open breech firing mechanism with bullet recovery included in the design, such as shown schematically in Figure 22.4, with volumes of both sampling areas to be kept to a minimum.

It would also be desirable to investigate the use of acidified potassium permanganate as a means of absorbing the mercury vapor, which can then be titrated with dithizone for quantitative analysis.

To finalize the series of tests on the distribution of mercury a further test was conducted to determine the amount of mercury deposited on the firer. The results are given in Table 22.14. The results demonstrate the random nature of FDR deposition; however, the firing hand was positive in all firings.

Table 22.14 Amount of Mercury Deposited on Firer

	Mercury (ng)				
No. of Shots	Right Hand	Left Hand	Face	Hair	Clothing
1	87	None	None	None	None
1	40	None	22	None	70
3	56	None	20	None	68
3	25	None	None	None	79
7	66	None	29	None	88
7	170	11	42	44	60

Table 22.15 Loss of Mercury with Time

Time Interval	Weight Range µg Hg	Average µg Hg
Unfired	3,813–5,938	4,952
Immediate	710–860	787
3 days	655–905	789
10 days	710–890	780
69 days	444–631	537

An ancillary test involved examining discharge residue particles from the ammunition used in the mercury distribution tests. Approximately 260 particles were examined of which 8 contained mercury (about 3%) at minor or trace level.

This number of particles is unlikely to account for the levels of mercury on the firing hand in the previous test. The possibility that some non-SEM detectable mercury is deposited on the firer cannot be excluded.

Loss of Mercury from Spent Cartridge Cases

The rate of loss of mercury from spent cartridge cases was examined to determine if this would assist in estimating the time of discharge. Results are presented in Table 22.15. As can be seen the rate of loss of mercury is too slow to be of practical value in estimating time of discharge. Again, there is wide variation in the amount of mercury initially present which is reflected in the variation in the amount remaining in the spent cartridge case. The amount remaining in the spent cartridge case immediately after firing (about 16%) is roughly in agreement with that determined in the distribution experiment (about 13%). The slow rate of loss of mercury (about 32% in 69 days) coupled with the wide variation in amounts initially present excludes the mercury level as a useful indicator of the time of discharge.

A further test involved storing spent cartridge cases for 3 days at 80°C. The elevated temperature greatly increased the rate of loss of mercury; the weight range was 351 to 432 µg, the average 378 µg. This suggests that climatic conditions (temperature) would be among the many factors involved in the rate of loss of mercury from the spent cartridge case.

References

200. G. M. Wolten, R. S. Nesbitt, A. R. Calloway, G. L. Lopel, and P. F. Jones, "Final Report on Particle Analysis for Gunshot Residue Detection," The Aerospace Corporation, El Segundo, CA. Aerospace report no. ATR-77 (7915)-3 (September 1977).
201. A. Zeichner, N. Levin, and M. Dvorachek, "Gunshot Residue Particles Formed by Using Ammunitions That Have Mercury Fulminate Based Primers," *Journal of Forensic Sciences* 37, no. 6 (November 1992): 1567.

Lead-Free Ammunition 23

Airborne lead, in both particulate and vapor form, arising from the discharge of ammunition, can be a health hazard to those regularly exposed over a long period of time. People, such as firearms training instructors, are at risk as lead has a cumulative effect in the body, eventually leading to serious illness. Antimony and barium are also undesirable from a health viewpoint, but their levels in discharging residue are considerably lower than the lead level. Despite elaborate indoor range extraction and ventilation systems, which do much to reduce the risk, continuous exposure over the years is undesirable and many organizations monitor blood lead levels (and hearing) of staff involved.

The first serious attempt to solve the problem at source was in 1978 when Smith & Wesson encased the entire lead bullet in a coating of black nylon (Nyclad ammunition). This substantially reduced the amount of lead released into the atmosphere from the discharge process, greater than 60% reduction when compared to the "cleanest" of conventional lead bulleted ammunition. As mentioned previously, most of the lead originates from the bullet, the lead in the primer making a much smaller contribution.

Analysis of a round of .38 Special caliber, Smith & Wesson Nyclad ammunition revealed that the cartridge case and primer cup are nickel-plated brass, no inorganic additives to the propellant were detected, the primer composition contains lead, antimony, and barium, the bullet core is antimony-hardened lead, and the bullet jacket contains traces of calcium, cobalt, titanium, and phosphorus.

The Northern Ireland Forensic Science Laboratory (NIFSL) was involved in the assessment of such ammunition for the police, and tests were conducted to compare lead levels using Nyclad and conventional ammunition. Unfortunately, the tests results did not survive the terrorist explosion at the laboratory in September 1992, but the overall conclusion was that the claims made relating to the reduction of lead were more than justified. (Other tests involving the examination of the perimeter of bullet holes in cloth that were caused by Nyclad bullets revealed that cobalt was repeatedly and readily detected by FAAS.)

The next stage in the development of environmentally friendly ammunition was when Geco produced ammunition with a totally jacketed bullet (TMJ) and a primer composition containing much reduced lead, antimony,

Table 23.1 Geco Primer Analysis

Ammunition Details	Pb µg	Sb µg	Ba µg
Geco .38" Special, metal piercing bullet with enclosed base	4.2	1.4	0.06
Geco .357" Magnum, metal piercing bullet with exposed base	91.5	130.0	4.4

and barium levels. An example of such ammunition was compared with its nearest equivalent to indirectly assess its effectiveness. Results are given in Table 23.1.

The .38" Special ammunition with the bullet base enclosed coupled with the substantially lower levels of lead, antimony, and barium in the primer will obviously give much lower levels of lead, antimony, and barium in the discharge products than the .357" Magnum caliber ammunition.

The nylon-coated bullets from Smith & Wesson and the Geco TMJ bullet/new primer composition, although effective in markedly reducing the lead levels, did not totally eliminate the problem. In 1983, Dynamit Nobel introduced 9 mmP caliber ammunition with a TMJ bullet and a primer free of lead, antimony, barium. The new primer type was called Sintox. A typical Sintox primer composition contains 15% diazodinitrophenol (DDNP) and 3% tetracene as the "explosive" ingredients, 50% zinc peroxide as the oxidizer, 5% of 40-µm size titanium metal powder, and 27% nitrocellulose as propellant powder.[202] As expected, other munitions manufacturers eventually introduced similar ammunition, some with primers that were lead free but containing antimony and barium and others free of lead, antimony, and barium. The objective was to produce ammunition which performed satisfactorily in every way and did not produce any toxic product on discharge.

A selection of "lead-free" ammunition was dismantled and analyzed and the results are given in Table 23.2.

The analysis of the organic components in the primers could suffer contamination from the propellant, although the results suggest that this is not the case. Dinitrophenol is probably a breakdown product of DDNP arising from the conditions used in GC/MS.

Environmentally friendly ammunition was initially introduced for training purposes in indoor firing ranges, but it is now gaining acceptance as conventional ammunition.[202] The use of such ammunition has yet to be encountered in casework in Northern Ireland but it is highly likely that it will be in the future. This will pose major problems for existing methodology in the characterization of discharge particles by the particle analysis method, in range of fire determination, and in the identification of bullet holes. Some workers are already addressing the problems.[202–207]

Anticipating its use in casework it was decided to look at the nature of discharge particles from lead-free ammunition. Typical results are presented in Table 23.3.

Lead-Free Ammunition

Table 23.2 Analysis of Lead-Free Ammunition

Ammunition	Cartridge Case	Primer Cup	Bullet	Propellant	Primer
(1) 9 mmP IVI 92 ⊗ Greenshield TM	Brass	Ni-plated brass	Frangible–powdered Cu in caprolactum binder; a number of alkane nitriles and long-chain alcohols also detected	NG, EC, trace DPA; K, S also detected	Sb, Ba, S (trace Fe, K, Cl, Si, Cu, Zn, Na, Ca); dinitrophenol
(2) CCI 9 mm Luger Speer–Lawman CF	Brass (trace Fe)	Ni-plated brass	Lead core (trace Cu, Fe, Sb) total Cu jacket	NG, EC, trace DPA; K, S also detected	Sr, Ti, S (trace Ba, Fe, Ca, K, Cu, Zn, Cl); dinitrophenol, EC, DPA (trace DNT, a nitro DPA, a phthalate plasticizer)
(3) CCI NR 9 mm Luger Blazer	Al (trace Cu, Mn)	Ni-plated brass	Lead core (trace Al, Fe) total Cu jacket	NG, EC, trace DPA; K, S also detected	Sr, S, Al, Fe (trace Ca, Cl, Ti, Cu, Si, Mg, Zn); EC, DPA (trace DNT, a phthalate plasticizer, a nitro DPA, benzyl butyl phthalate plasticizer)
(4) GFL 9 mm Luger Fiocchi	Brass	Brass	Lead core, Al base cup, full brass jacket	NG, EC, K, Cl (trace Cu, Zn, Cl, K, Ca, Fe, S, Ti, Mg, P) also detected	Sb, Ba, S (trace Al, Si, Fe, Cu); dinitrophenol
(5) ◊ — ◊ 9 mm Luger Delta	Brass (trace Al)	Ni-plated brass	Frangible–powdered Cu with a trace of W in caprolactum binder; a number of alkane nitriles also detected	NG, EC, trace DPA; K, S also detected	Sb, Ba, S (trace W, Cu, Ca, Fe, Si, Zn, Al, Mg); dinitrophenol, NG and a number of N-alkanes

Table 23.2 Analysis of Lead-Free Ammunition (Continued)

Ammunition	Cartridge Case	Primer Cup	Bullet	Propellant	Primer
(6) Sintox 9 mm Luger	Brass	Ni-plated brass (trace Fe)	Lead core (trace Fe, Cu, Sb), brass (trace Fe) base cup, full brass-coated Fe jacket (trace Al)	NG, DPA, a phthalate plasticizer (trace DNT and a nitro DPA); K, Ca (trace S) also detected	Zn, Ti (minor S, trace Cl, Al, Cu, Ca, Fe, K, Si, Na); dinitrophenol, a methyl propyl phenol, DPA, a phthalate plasticizer, a trace of DNT
(7) 9 × 19 SX DAG 88	Brass (trace Al)	Ni-plated brass	Solid brass with a plastic filled hollow point nose	NG, DPA, a nitro DPA, a phthalate plasticizer (trace DNT and other phthalate plasticizers); K, trace S also detected	Zn, Ti (trace Cu, Ca, K, S, Fe, Si); NG, DPA, a phthalate plasticizer, a methyl propyl phenol, trace DNT
(8) WIN 9 mm Luger Delta	Brass	Ni-plated brass	Frangible-powdered Cu with minor W in caprolactum binder; a number of alkane nitriles also detected	NG, EC, trace DPA; K, S also detected	Sb, Ba, S (trace Fe); dinitrophenol (trace DPA, EC)
(9) CCI NR 38 SPL +P Blazer	Al (trace Cu, Mn)	Ni-plated brass	Lead core (trace Fe), total Cu jacket	NG, EC, trace DPA; K, S also detected	Sr (trace Na, S, Al, Ca, K, Ti); dinitrophenol, EC, DPA (trace DNT, 2 × nitro DPA, a phthalate plasticizer)
(10) WIN 9 mm LugerWinchester Super X	Brass (trace Fe)	Ni-plated brass (trace Fe)	Lead core (trace Al, Fe), brass (trace Fe) base plate; full Cu jacket (trace Fe, Zn)	NG, DPA, a nitro DPA, a phthalate plasticizer (trace DNT and EC); K, S, and Ca also detected	Zn, Ti (minor S, Ca, K, Si, Mg, Al, trace Fe, Cu); a series of N-alkanes with phthalate plasticizers

(11) M&S 38 Special	Brass (trace Al)	Brass	Frangible–powdered Cu, Al in caprolactum binder; a number of alkane nitriles also detected	NG, EC, 2 × nitro DPA (trace DNT, glycol dilaurate plasticizer); K, trace S also detected	Pb, Sb, Ba, Al (trace S, Ca, Cu, Fe, Si); NG and a number of N-alkanes
(12) .223 IVI ⊕ 91	Brass (trace Fe, Al)	Ni-plated brass (trace Fe, Al)	Frangible–powdered Cu, Al in caprolactum binder; a number of alkane nitriles and long-chain alcohols also detected	NG, DPA, a phthalate plasticizer with trace EC, DNT, a nitro DPA; K, S, Ca also detected	Sb, Ba, S, Ca (trace Al, Cl, Cu, K, Fe); NG, a phthalate plasticizer and a number of N-alkanes
(13) HP .223 5.56 CF	Brass (trace Fe)	Ni-plated brass	Plastic core, full Fe jacket, copper coated	NG, DPA, a phthalate plasticizer (trace DNT, EC, a nitro DPA); Ca, K, S, Cl also detected	Sr (minor Ti, trace Cu, Ca, S, Zn, K, Fe); dinitrophenol, DPA, EC, A phthalate plasticizer (trace DNT, methyl propyl phenol)

Note: See Glossary for firearms/ammunition-related abbreviations.

Table 23.3 Discharge Particle Types from Lead-Free Ammunition

Ammunition	Typical Particle Types			Comment
	Major	Minor	Trace	
Nos. 1, 4, 5, 8, and 12 Table 23.2 Sb, Ba primed	S, Sb, Ba, Cu	—	K, Cl, Si	Size range 1–14 µm
	S, Sb, Ba	—	K, Cl, Si, Cu	
	Cu, Zn	S	K, Sb, Ba, Si	Large number of Sb, Ba particles
	S, Cl, Sb, Ba	Cu	Si, Na	
	Ba, S, Cl	—	Cu, K, Si, Ca	No. 5 produced a number of particles containing tungsten
	Sb, S	—	Cu	
	Ba	Sb, S	Si	
	Sb, Ba	S	Cu	Mainly irregular particles
	Sb	—	—	
	Cu, S, Ba, Sb	Zn, Si	Fe	
Nos. 2, 3, 9, and 13 Table 23.2 Sr, Ti primed	Sr	S, Ca	Ti, Cu, K	Size range 1–10 µm
	Sr, Ti, S	Ca, K	—	
	Sr	—	S, Ca, K, Ti	Large number of Sr, Ti particles and Sr-only particles detected
	Sr, S	Ti, Cu	Zn	
	Cl, Sr, Ti	S, Cu, K	—	
	Sr, Al	—	—	
	Sr	—	Ca, K, S	A few Ti-only particles were noted
	Ca, Ti, Si, S	Zn	Cl, Cu, Al	Mixture of irregular, spherical, and oval particles
Nos. 6, 7, and 10 Table 23.2 Zn, Ti primed	S, Zn	—	Cl, Al, Cu	Size range 1–6 µm
	Zn, Ti	Ca	Cu, S	
	Ti, Ca	Zn	S, Cu, Si	Large number of Zn, Ti particles
	Ti	—	Zn, Ca, K	
	Zn, Ti	S, Al	—	A substantial proportion of Zn-only and Ti-only were detected
	Zn, Ti	—	S, Na	
	Ti	Ca	—	
	Zn	Ti	Ca	
	Zn	—	—	Mostly spherical and oval particles
	Zn	—	Cu	

Ammunition with primers containing antimony, barium should not pose a problem, as discharge particles produced contain antimony and barium, particles classified as unique to the discharge of a cartridge. However, the other primer types will cause problems. A study of the discharge particles from Sintox ammunition concluded that morphology is an essential identification criterion and that spherical particles composed mainly of titanium and zinc can be used to identify discharge particles from Sintox ammunition.[202] (It is reported that Dynamit Nobel has replaced calcium silicide with titanium in Sinoxid type primers. Consequently, titanium as well as lead, antimony, and barium may be encountered in discharge particles.)

It would be desirable to examine environmental and occupational particles containing one or more of the elements strontium, titanium, and zinc and to compare them with discharge particles, in order to establish whether or not it is possible to differentiate the FDR particles. If not, the particle analysis method may still be a useful tool to provide supporting evidence, particularly if organic propellant constituents are also detected on the suspect.

References

202. L. Gunaratnam, and K. Himberg, "The Identification of Gunshot Residue Particles from Lead-Free Sintox Ammunition," *Journal of Forensic Sciences* 39, no. 2 (March 1994): 532.
203. R. Beijer, "Experiences with Zincon, a Useful Reagent for the Determination of Firing Range with Respect to Leadfree Ammunition," *Journal of Forensic Sciences* 39, no. 4 (July 1994): 981.
204. G. M. Lawrence, "Lead-Free or Clean-Fire," *Southwestern Association of Forensic Scientists Journal* 15, no. 2 (October 1993): 44.
205. W. Lichtenberg, "Examination of the Powder Smoke of Ammunition with Lead-Free Priming Compositions," *Kriminalistik* (December 1983): 377.
206. "Examination of the Powder Smoke of a Recent Type of Ammunition with Conventional Priming Composition and Changed Projectile," Paper 9, *Proceedings of the Conference on Smoke Traces,* Bundeskriminalamt (Federal Criminal Investigation Office), Wiesbaden, June 1985.
207. W. Lichtenberg, "Study of the Powder Smoke of the CCI, Blazer, cal .38 Ammunition, by the Film Transfer Method and X-ray Fluorescence Analysis," *KT-Material—Information*, no. 4 (October 1986): 4.

VI

Suspect Processing Procedures

Firearm Discharge Residue Sampling

24

The sequence of events associated with FDR examination is usually as follows: the initial incident, apprehension of suspects, transporting of suspects to police station, sampling of suspects at police station (swabs of hands, face, head hair, and seizure of clothing), submission of items to the laboratory, sampling of clothing at the laboratory, sample preparation, analysis of samples, interpretation of results, preparation of statement of witness report, and the presentation of forensic evidence in court.

With any process involving sampling, if the sampling is poor or incorrect, then all subsequent manipulations, observations, and conclusions can be devalued or meaningless. It was decided to review the complete process, starting with the processing of suspects. (The review was done prior to the development of a method for organic FDR detection, in the hope that improvements could be identified and implemented in suspect processing procedures, thereby increasing the chances of obtaining positive results in most areas of forensic work, including the existing inorganic FDR system and any new organic FDR system.) Procedures for dealing with suspects from the time of apprehension to the time of forensic sampling were considered with a view to identifying any weaknesses in the system that could result in contamination and/or loss of any type of forensic evidence. Recommendations designed to eliminate those weaknesses should then, if implemented, increase the chances of obtaining forensic evidence and/or strengthen the value of evidence obtained.

There are two very important factors involved in the processing of suspects for forensic evidence:

1. Time delays which substantially diminish the chances of obtaining evidence.
2. Contamination risks which substantially diminish the value of evidence obtained.

Time Delays

Time delays greatly reduce the chances of obtaining positive results, a fact that is well documented in the scientific literature, for example, references

208 and 209. This applies to a greater or lesser extent to nearly all types of forensic evidence originating from the examination of suspects. Firearm and explosive residue in particular are lost very rapidly with the passage of time (90% of firearm residue is lost from the hands in the first hour after firing, even during the course of normal activity). Suspects must be sampled as a matter of urgency.

Reasons for delays include lack of availability of trained samplers and materials, organizational problems, and geographical location.

Large time delays can occur in certain cases because of geographical locations. Suspects should be taken to the nearest sampling location irrespective of divisional boundaries. After sampling they could then be transferred to the appropriate premises if necessary. Although this may generate extra paperwork and inconvenience, obtaining evidence should be the paramount consideration.

At each sampling location there should be trained personnel on site at all times. The more trained personnel available at any location, the greater the chances of reducing time delays to a minimum.

At each sampling location all the necessary sampling materials should be readily available at all times. A clean storage area should be provided at each sampling location for storing materials required for the forensic sampling of suspects. This would avoid the possibility of kits or packaging materials becoming contaminated while stored at other locations or carried in vehicles.

A conveniently situated, modest-sized (approximately 120-square-foot) storage area would suffice and it should be smooth, light colored, linoleum floored with Formica type shelving, non-heated, and dry. (Bear in mind that some sampling kits contain flammable solvents.) The store should be lockable and under the directed control of an appointed officer and used *exclusively* for the storage of suspect sampling materials for use at that location only. There must be a strict accounting system for all materials.

This is the most obvious solution to the time problem. Having fully trained samplers permanently on site, 24 hours a day every day, coupled with all the necessary sampling materials permanently on site, is by far the most sensible and effective way of solving many of the time delays.

Contamination Avoidance

The major problem associated with trace evidence is the possibility of cross transfer to the suspect from some unrelated source. In Northern Ireland there must be a greater contamination risk than in the rest of the United Kingdom, due to the relative abundance of firearms and explosives. Cross-contamination allegations are a frequently used defense in court. Contamination risks have been exaggerated out of all proportion, and all problems in this area stem from the difficulty in providing basic facts, knowledge, and statistics

Firearm Discharge Residue Sampling

to demonstrate that cross contamination is unlikely to have occurred. This statement is based on court experience and the fact that the vast majority of suspects examined for firearm and explosive residue are negative. If contamination is such a problem it is certainly not reflected in laboratory results. None of this is meant to suggest complacency. These "negative" suspects are ideal "controls" and consequently provide the "ammunition" with which to hit back at the frequently used defense of cross contamination. This defense is very often invalid but hard to refute completely. To resolve this situation accurate casework statistics on computer should be compiled for use in court.

In multisuspect cases the suspects should be separated immediately and remain separated at all times. Suspects are apprehended and conveyed to the sampling location in police vehicles. Ideally, a suspect should go directly to a room and be under constant supervision until sampling is complete.

It is most important that all suspects be kept separated and supervised at all times between arrest and sampling. On arrival at the sampling location they should individually be taken directly to a room and be under constant supervision until sampling is completed. This would avoid the risk of cross contamination during the processing stage. A number of "clean" rooms should be provided at each sampling location and the room should be cleaned between suspects. Control samples should be taken by laboratory staff at frequent intervals to monitor the cleanliness of the rooms. The materials storage area, cells, custody offices, medical examination rooms, and police vehicle interiors should also be monitored at frequent intervals.

Evidence Protection Kit

One potential major improvement in suspect handling lies in the use of a kit, specifically designed to retain evidence on a suspect. It has for a long time been a source of annoyance and frustration that, at the time of apprehension of a suspect, there are currently no effective measures available to secure potentially crucial evidence. Some police forces throughout the world use bags, usually plastic ones, to cover the hands of persons suspected of firing a gun. The use of plastic bags can have a detrimental effect on FDR, causing the hands to sweat, which in turn causes most of the FDR to be removed from the hands onto the surface of the bag. Apart from protection of the hands, no effort has been made to protect other suspect sampling sites.

Existing systems fail to prevent the loss of evidence from the most significant areas during the period between apprehension and sampling, as well as providing many opportunities for cross contamination to occur. If one could prevent the loss of evidence and confine any contamination risk to the initial arrest procedures, the advantages would be enormous.

A kit for the protection of a suspect's hands before sampling for firearm and/or explosive residue has been used in Northern Ireland for many years. The kit involved placing the suspect's hands inside paper bags to prevent the removal of residue prior to sampling. In use it was cumbersome, because the bags have to be applied before securing the suspect's arms. Unfortunately, suspects are rarely apprehended until several hours after the incident, and because of the rate of loss of such traces from the hands, the success rate was very low. The suspect's upper outer garment proved to be a much more fruitful area for the detection of firearm and explosive residue. This led to some thought being directed toward a means of protecting both the upper outer garment and hands prior to sampling, in an effort to reduce the rate of loss of all types of forensic evidence. Consequently, a kit for use by the police during apprehension of the suspect was designed and commercially produced, and has been operational since early 1993 without problems. Details of the kit have been published.[210]

Comment

The single most important factor in the avoidance of contamination problems is good housekeeping. This applies to all aspects of trace residue examination procedures, and includes police vehicle interiors, custody offices, cells, sampling rooms, and sampler's premises.

It is highly desirable for a forensic science laboratory engaged in such work to frequently monitor the sampling procedures and sampling environment used by the police in order to be in a position to demonstrate the validity of laboratory results. This would involve an ongoing commitment of laboratory resources, such as regular monitoring of sampling locations, samplers premises, police vehicles, kit making premises and products, for the presence of FDR and explosive residue, in addition to compilation of computer records of all casework involving the examination of suspects for either FDR or explosive residue.

The benefit is the ability to effectively refute cross-examination allegations in court, thereby increasing the confidence level of everyone involved, in both the laboratory results and interpretations based on the results.

A further improvement in contamination avoidance, instigated by the police, is the compulsory use of a contamination avoidance kit by anyone entering the scene of a crime. I was approached to design such a kit and my initial thoughts were: a suitable sized nylon bag divided into two sections by the use of a heat sealer, the first (upper) section containing instructions, an X-ray medi-wipe, a pair of large size disposable plastic gloves, and a face mask. The second (lower) section containing a large size disposable overall with integral hood, and overshoes. A demonstration kit was prepared and the design is currently being used by the police.

In conclusion, it is my opinion that the laboratory results should not be considered in isolation but should be interpreted with the knowledge that established and reliable contamination avoidance procedures are in operation and that they are monitored on a regular basis.

References

208. R. Cornelis, and J. Timperman, "Gunfiring Detection Method Based on Sb, Ba, Pb, and Hg Deposits on Hands. Evaluation for the Credibility of the Test," *Medicine, Science, and the Law* 14, no. 2 (April 1974): 98.
209. J. W. Kilty, "Activity after Shooting and Its Effect on the Retention of Primer Residue," *Journal of Forensic Sciences* 20, no. 2 (1975): 219.
210. J. S. Wallace, "Evidence Protection Kit," *Science & Justice* 35, no. 1 (1995): 11.

VII

Organic Components of Firearm Discharge Residue

Sampling of Skin and Clothing Surfaces for Firearm Discharge Residue

25

Introduction

There are several reasons the detection of organic constituents of FDR was worthy of investigation. The particle analysis method is tedious and the speed of analysis is very slow, thereby often not meeting all the needs of the police, who frequently require a fast answer at an early stage in an investigation. The detection of organics is much faster, particularly for multisuspect cases involving a large number of samples, and it was hoped that the time problem could be partly solved by screening for organics prior to inorganic examination, enabling preliminary results to be available for the police investigation.

If the organic system was as sensitive, if not more so, than the inorganic system and if there was a very good correlation between the detection of organic and inorganic FDR, then it would be sensible to use organic analysis to determine which samples are worthy of lengthy SEM/EDX examination. The vast majority of samples examined by the laboratory are negative for inorganic FDR. The screening technique is automated SEM/EDX,[211,212] which is a lengthy procedure. Positive samples are confirmed by manual examination, also a lengthy procedure.

The majority of terrorist shooting incidents in Northern Ireland involve the use of single-based propellants, and as the main thrust to date in organic FDR analysis had been based on the detection of NG from double-based propellants, using techniques such as GC/TEA and HPLC/PMDE, it was obvious that GC/MS was the instrumentation most likely to satisfy the requirement of detecting constituents from single-based propellant. From the start it was realized that the chief problem was the low concentration of detectable constituents present in single-based propellants which, after combustion, would be present at extremely low levels in the discharge residue. To increase the chances of success it was decided to optimize all aspects of the system, starting with decreasing the suspect sampling time by the police and increasing the efficiency of sampling techniques used by the police and the laboratory. From the outset, the work was planned so that all subsequent development work could be based on optimized and standardized sampling techniques.

The ability to detect organic constituents in FDR from ammunition with single-based propellant was the ultimate goal. If this proved impossible, the investigation would at least clarify the situation and improve the detection method for NG, which could serve as a useful complementary technique to SEM/EDX. If particles in the "indicative" category were accompanied by NG, it would substantially raise the significance level of such particles.

In the NIFSL two separate sections dealt with the detection of firearm and explosive residue. The trace analysis work of the firearms section and the explosives section are now done in one laboratory, called the microchemistry laboratory. Each section had two sampling kits, one for the sampling of suspects and one for the sampling of miscellaneous items, and each section had a different method for sampling clothing from suspects. It was decided to look at the possibility of using the same kit to sample for both firearm and explosive residue and to investigate the feasibility of a joint procedure for sampling clothing. Both the kit and clothing sampling procedures would have to be compatible with existing residue detection systems, primarily organic in the case of explosives, and inorganic in the case of firearms. It was also necessary to ensure that the organic components of FDR could be accommodated using existing instrumentation and that a single sample could be analyzed for both firearm and explosive residue. In the past, separate samples would have been taken, one for FDR and one for explosive residue. The desirability of routinely screening all terrorist suspects for both FDR (inorganic and organic) and explosive residue was also borne in mind.

A joint sampling procedure for clothing was devised and the four previous sampling kits were reduced to two, a joint kit for suspects and one for miscellaneous items.

Sampling of Clothing

In this part, the development of a clothing sampling method, common to both explosive and firearm residue is described.

Previous experiments on clothing demonstrated that suction sampling is considerably more efficient for the removal of FDR than either adhesive lifting or swabbing techniques. The experiments were conducted by repeated sampling of FDR-contaminated clothing, with the samples then analyzed for lead, antimony, and barium by FAAS. Recovery using adhesive tape was ~20% clothing ~37% hands; swabbing was ~25% clothing ~70% hands, and suction sampling was ~65% clothing (hands not examined). (Suction sampling is also suitable for recovery of FDR from head hair.) As a consequence the NIFSL has, since 1979, been suction sampling clothing for FDR using the apparatus in Figure 25.1 attached to an Edwards E2M6 vacuum pump (pumping rate = 108 liters per minute).

Sampling of Skin and Clothing Surfaces

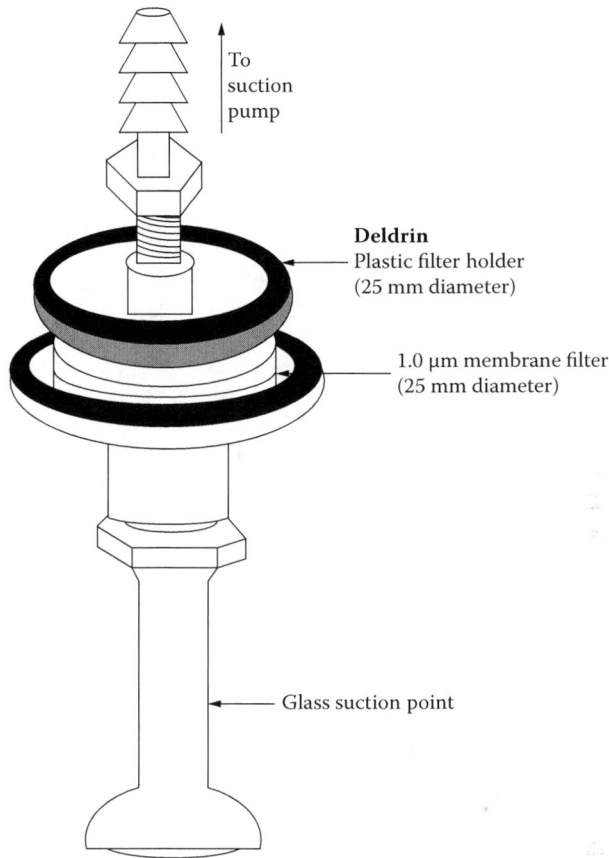

Figure 25.1 Suction sampling device.

Previously, when a garment was to be examined for firearm and explosive residue, certain areas were swabbed with cotton wool damped with acetone to collect explosive residue and other areas were suction sampled for FDR. It was decided to investigate the possibility of using suction sampling for both types of residue and to look at the existing suction sampling system with a view to improving its efficiency. As many of the propellant components are explosive or explosive-related compounds, it is a reasonable assumption that experimental results with organic FDR would also apply to explosive residue. A series of experiments was devised to compare and test clothing sampling procedures.

The first step was to directly compare existing procedures, namely, suction sampling and swabbing. Table 25.1 gives comparison data. The results reveal that NG was detected on all three swab samples and that it was not present on the third suction sample. The initial suction sampling removes a high proportion of the NG whereas the second swabbing shows a substantial level of NG (~40%) when compared to the initial swab sample. Suction

Table 25.1 Recovery of Organic FDR from Clothing

Sample	NG (ng)	2,4-DNT (ng)
Swab 1	272	ND
Swab 2	110	ND
Swab 3	26	ND
Membrane 1	876	17
Membrane 2	58	ND
Membrane 3	ND	ND

ND = not detected.

sampling detected 2,4-DNT, admittedly at a low level, but swabbing failed to reveal its presence.

It was concluded that suction sampling is more efficient than swabbing and the results suggest that suction sampling may also be suitable for explosive residue sampling. In order to confirm this, a further test was conducted involving doping a new laboratory coat with a mixture of explosive compounds and then suction sampling it. The mixture contained a known concentration of RDX, PETN, EGDN, NG, NB, TNT, 2,3-DNT, 2,4-DNT, 2,6-DNT, DPA, EC, and MC and was sprayed on to the garment and the solvent allowed to evaporate before sampling. The suction sample was submitted to the explosives laboratory for analysis using a combination of GC/TEA, HPLC/PMDE, and GC/MS techniques. The explosives section concluded that suction sampling is suitable for the recovery of explosives residue from clothing.[213]

As suction sampling is superior to other sampling techniques, such as adhesive lifting and swabbing, it was decided to try and optimize the existing suction sampling system. To this end the effect of the membrane pore size on the efficiency of recovery of NG was investigated using the apparatus shown in Figure 25.2. Results are given in Table 25.2.

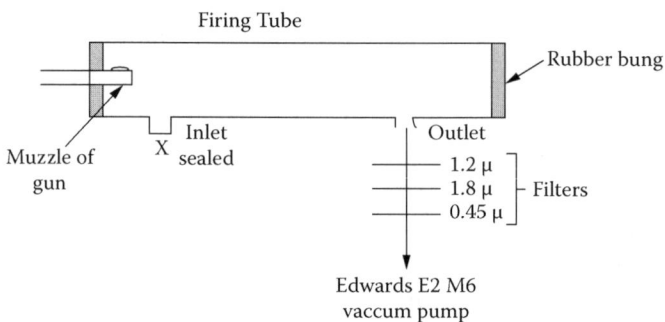

Figure 25.2 Apparatus for sampling discharge residue.

Table 25.2 Membrane Pore Size Experiment on Muzzle Residue

Test No.	Pore Size (µm)	NG (ng)
1	1.2	8,600
	0.8	355
	0.45	17
2	0.8	7,400
	0.45	200
3	1.2	9,050
	0.8	570
4	1.2	8,100
	0.45	315

Although the results were not definitive, it is clear that substantial amounts of NG passed through both the 1.2- and 0.8-µm filters, although the initial filter retained the bulk in each test. (There were no membrane filters with a pore size less than 0.45 µm in the laboratory at the time of the test. It would have been interesting to repeat the experiment incorporating the smaller pore size filters, such as 0.2 µm.) It is suspected that NG is present as both vapor and particulate matter, the particulate matter having a wide size range.

In an attempt to clarify the situation and to validate the system the experiment was repeated on clothing worn while firing. Results are given in Table 25.3 and Table 25.4.

As the 1.2-µm and the 0.8-µm filters allowed a significant amount of NG through, it was decided to introduce the 0.45-µm filter coupled with a more powerful pump (Edwards E2 M12: pumping rate ~240 liters per minute), the increase in pump capacity to enable two suction lines to be run to two

Table 25.3 Membrane Pore Size Experiment on Clothing (one shot)

Test No.	Pore Size (µm)	NG (ng)
1	1.2	100
	0.8	ND
	0.45	ND
2	0.8	160
	0.45	5
3	1.2	90
	0.8	3
4	1.2	200
	0.45	20

ND = not detected.

Table 25.4 Membrane Pore Size Experiment on Clothing (six shot)

Test No.	Pore Size (μm)	NG (ng)
1	1.2	320
	0.8	5
	0.45	ND
2	0.8	480
	0.45	15
3	1.2	270
	0.8	8
4	1.2	335
	0.45	17

ND = not detected.

sampling areas, thereby improving our sampling facilities. (Later experience proved the 0.45-μm MCE membrane to be affected by the solvents used in the organic FDR procedure and it was changed to a 0.5-μm PTFE membrane, which has since proved to be entirely satisfactory.)

The glass disposable spout on the suction sampling apparatus was originally introduced to facilitate the sampling of pocket interiors, but with experience it was decided that the glass spout was not necessary as it restricted the suction area, reduced the efficiency of removal, and increased the time taken to sample a garment. Consequently the revised sampling device is as shown in Figure 25.3.

The revised suction sampling device was tested by comparison with the previous device. Results are presented in Table 25.5. This represents a good recovery of NG when compared to Table 25.4 and illustrates that the device is effective. However, in this instance only one membrane filter was present in the suction line; consequently, like is not being compared with like. The improved recovery may be due to a combination of factors.

A final test was conducted, using everyday clothing, to simulate a "real-life" situation. The test was designed to verify the superiority of suction sampling over swabbing and also to check that the entire system, including inorganic particle detection, was satisfactory. The test involved using both sampling techniques on the same FDR-contaminated garment and alternating the order of sampling. Results are given in Table 25.6.

The results show that suction sampling removes nearly all the detectable NG present whereas swabbing removes considerably less. The swabbing prior to suction sampling probably had a detrimental effect on the number of inorganic particles recovered.

Sampling of Skin and Clothing Surfaces

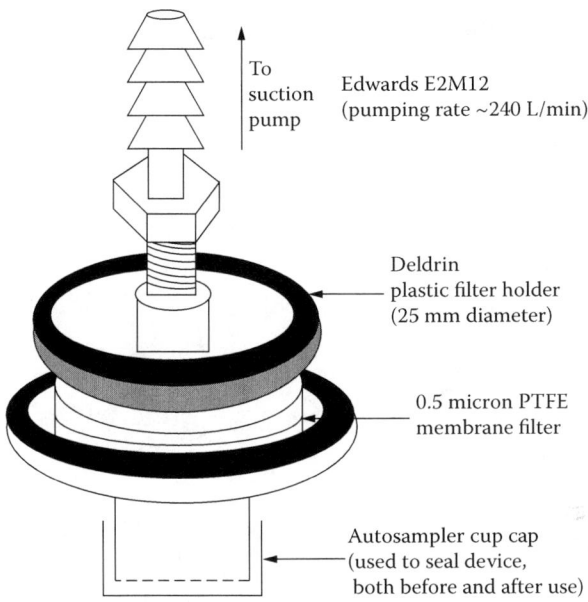

Figure 25.3 Revised suction sampling device.

Sampling Kits

The previous method for sampling a suspect's skin surfaces for explosive residue was swabbing using cotton wool damped with acetone, and the previous firearms method involved swabbing with acrilan fiber damped with petroleum ether. The two previous general-purpose kits for sampling miscellaneous items used the same techniques and materials, except that dry swabbing was used in the firearms kit. When a suspect had to be examined for both types of residue, swabs from the palm of the hand plus nail scrapings were taken for explosive residue and the backs of the hands were swabbed for FDR, as was the face and head hair. This was necessary as the use of one sampling technique excluded the subsequent use of the other on the same area, and caused confusion and considerable inconvenience as two types of kit had to be used to sample one suspect. The firearms kit was incompatible

Table 25.5 NG Recovered from Clothing (six shot)

Test No.	NG (ng)
1	720
2	660
3	670
4	615

Table 25.6 Comparison of Swab and Suction Sampling Techniques

	Sampling Procedure	NG (ng)		Inorganic Particles (suction sample)			
		Suction	Swabbing	Pb, Sb, Ba	Sb, Ba	Pb, Sb	Pb, Ba
Jacket 1	Swabbed first, then suction sampled	1,080	495	8	2	8	7
Pullover 1	Swabbed first, then suction sampled	915	810	13	1	24	2
Jacket 2	Suction sampled first, then swabbed	1,555	ND	27	None	16	None
Pullover 2	Suction sampled first, then swabbed	2,670	Trace	15	None	7	4

ND = not detected.

with laboratory preparation procedures for explosive residue, and vice versa. Similar problems were encountered in the sampling of clothing.

The analytical methods currently used by this laboratory are chromatography (GC/TEA: HPLC/PDME) for explosive residues and the particle analysis method (SEM/EDX) for FDR. The latter method involves the detection and identification of individual FDR particles; therefore any sampling technique must be nondestructive.

At the kit design stage many factors had to be considered, including lifting efficiency, compatibility with existing laboratory techniques, the avoidance of cross contamination between sampling areas and possible contamination from other sources, for example, the sampler and the sampling room. Ease of use and preparation, the cost, purity and availability of materials, and the shelf life are all important, as well as the health and safety aspects. Finally, it needs to be acceptable to courts of law and to other scientific examiners. Previous kits suffered a limited shelf life due to evaporation of the solvents. To solve this problem the new kits use a predamped acrilan swab in a sealed metal foil sachet. Joint sampling kits were designed, tested, and prepared in the laboratory before issue to the police. Details of the revised sampling kits and procedures have been published.[214]

Clothing Examination

As discussed previously, a single sampling technique for clothing that satisfies the requirements of both types of residue has now been devised. The

Sampling of Skin and Clothing Surfaces

entire outside front surface (one sample) and pocket interiors (one sample) of a suspect's upper outer garment is routinely sampled for firearm and/or explosive residues by laboratory staff. In addition, other areas and other garments may be sampled depending on the nature and circumstances of the case. Control samples are also taken, one from the outside of the nylon packaging bag containing the garment and a combined control from the sampler, the work surface, and the air in the sampling room. The apparatus currently used is shown in Figure 25.2.

It consists of a 25-mm-diameter Deldrin filter holder (Gelman product No. 1109) with one of the nylon hose nipples removed. The filter used is a 25-mm-diameter, 0.5-μm pore size Fluoropore membrane filter (Millipore catalogue No. FHLP 025 00). An auto sampler cup cap is used to seal the apparatus, both before and after use. In use, the filter holder is attached to an Edwards E2 M12 vacuum pump.

Discussion

At the initial stage, the suspect sampling kit was designed to be as self-contained and secure as practical. Everything needed for the procedure, including packaging and labeling materials for submission to the laboratory, is in the kit. This approach lessens the possibility of contamination as all materials involved are under the control of the laboratory, and it also makes the kits more user friendly. Security is enhanced by the issue of the kit in a sealed nylon bag, and by the swab sample container having a tamper-proof seal and the requirement for its return in a similar fashion. All swabs are in a sealed metal foil sachet and all swabs, used or unused, must be returned to the laboratory. All kits have a laboratory code number and logo on them and are only issued against a signature. As a consequence, laboratory records show how many kits are issued, when and to whom. Incoming kits are checked against the issue records. Integrity labels, included in the kit, are used initially to seal the nylon packaging bag used for submission of the kit to the laboratory.

At the design stage the assumption was made that both the sampler and sampling room were contaminated. The avoidance of the possibility of contamination is improved by the procedures used. The kit contains three control swabs, one of which is left untouched by the sampler and serves as a control of the kit materials, another serves as a control of the sampler, and the third serves as a control of the sampling environment. Cross contamination between sampling areas is avoided by the changing of the sampler's gloves between samples. Polythene gloves are used because other types of disposable gloves have a lubricant present which can contaminate the samples and cause problems during a search for FDR particles in the SEM. The kit is sealed inside a nylon bag which is less permeable than a plastic bag, and a quantity of kits are issued inside a larger sealed nylon bag from where they are removed only as required, after which the bag is resealed. The overall is worn by the sampler to avoid the possibility of any transfer of residue on the sampler's clothing to the suspect or work surface. The work surface is cleaned before use, and in addition, a plastic "tablecloth" is used as a clean work area.

The kit contains a detailed incident report form. The kit instructions are signed and dated by the sampler as a true and accurate record of the process, and are designed to be retained by the sampler as contemporaneous notes for use in court. The kit is suitable for both destructive and nondestructive testing, and can also be used for sampling suspects in cases involving the theft of metals.

The revised suspect sampling kit and general-purpose sampling kit have now been operational in casework for 12 years during which time no problems have been encountered, either in the use of the kits or in subsequent laboratory preparation and analysis procedures.

Suspect sampling time is in the region of 30 to 40 minutes, and the removal and packaging of the suspect's outer garments take a further 5 minutes. The kits are more costly and time-consuming to prepare than the equivalent previous kits, but as two new kits replace four previous kits, the overall cost and preparation time are less. None of the components of either kit is reused.

Precautions to prevent cross contamination, incorporated in the new suspect sampling kit, play a significant role in negating cross-contamination claims arising from the sampling procedure. Quality of the kits is controlled by strict preparation and packaging procedures. It could be necessary to demonstrate in court that all reasonable precautions have been taken during manufacture to avoid contamination occurring, and also that no evidence of contamination was found. To this end, there are clearly defined written procedures to be followed during preparation and packaging stages, including the use of hand-washing, disposable overalls, disposable gloves, and cleaning processes for certain components. Additionally, swab samples are taken daily from personnel, protective wear, and accommodation (tables, benches, chairs), and the completed kits are subjected to random sampling. Analytical results are retained for possible future use. A swab and a plastic tube are included in every kit as a kit (materials) control, and are left untouched by the sampler.

Suction sampling of clothing for FDR examination has been used by this laboratory since 1979. As discussed previously, the suction sampling apparatus described was modified from that previously used to increase its efficiency and to make it compatible with explosive residue examination. Such sampling is efficient and easy to use but suffers from the disadvantage that the Deldrin filter holder is too expensive to be regarded as disposable.

Consequently, a rigorous washing procedure is necessary to ensure no carryover of organic or inorganic residues. This does not pose a problem as all traces of firearm and explosive residues can be removed by rinsing under running water, soaking overnight, first, in 2% v/v Decon; second, in 2% v/v hydrochloric acid; and third, in 2% v/v bleach, followed by rinsing in deionized water and then in acetone. Apparatus blanks and control samples are routinely taken, whose results testify to the efficiency of the washing procedure. The overall result of unification of procedures is a modest saving of time and money, but, more importantly, both police and laboratory staff have benefited from the simplification of the system.

References

211. T. G. Kee, C. Beck, K. P. Doolan, and J. S. Wallace, "Computer Controlled SEM Micro Analysis and Particle Detection in the Northern Ireland Forensic Science Laboratory—A Preliminary Report," Home Office Internal Publication. Technical note no. Y 85 506.
212. T. G. Kee, and C. Beck, "Casework Assessment of an Automated Scanning Electron Microscope/Microanalysis System for the Detection of Firearms Discharge Particles," *Journal of the Forensic Science Society* 27 (1987): 321.
213. Private communication, 1994.
214. J. S. Wallace, and W. J. McKeown, "Sampling Procedures for Firearms and/or Explosives Residues," *Journal of the Forensic Science Society* 30 (1993): 107.

Development of a Method for Organic Firearm Discharge Residue Detection

26

The requirements of a method for detecting organic FDR were that it would not be detrimental to the existing particle analysis method, that the results would be meaningful, and that it should integrate into the two individual existing systems for the detection of explosive and firearm residue, thereby creating a system which could, if desired, routinely analyze all samples for organic explosives and organic and inorganic FDR. Because a large proportion of our casework involves the use of ammunition with single-based propellants, it was desirable to investigate the possibility of using GC/MS to detect constituents of propellant that are not detectable using our existing explosive residue analysis system. Such constituents include DPA, MC, EC, camphor, and phthalates. Because these constituents are originally present in relatively small amounts (typically 0.5% to 2.0%), compared to the level of NG in double-based propellants, it would be necessary to optimize all aspects of the system, namely, the extraction and cleanup/concentration procedures in addition to the instrumental parameters.

As stated previously, the existing techniques used for explosives residue detection of nitro compounds would be suitable for the detection of similar compounds in propellant. Initially the method of Lloyd[215,216] was employed, but it was found that the cleanup/concentration procedure was tedious, lengthy, and not robust enough for routine application with a large FDR caseload. Another major disadvantage was the back flushing procedure for the recovery of inorganic particles for SEM/EDX analysis. It was found that after back flushing, a substantial number of inorganic particles remained in the acrodisc which was used for suction sampling clothing. This was determined by cutting open acrodiscs after back flushing and examining their interior by SEM/EDX. As the main consideration in the development of an organic FDR method was that it must not be detrimental to the existing particle analysis method, the back flushing procedure was unacceptable, and it was therefore necessary to develop our own systems.[217]

Although there is extensive literature on the identification and detection of propellants, little work has been devoted to the combined analysis of organic and inorganic FDR recovered from hands and clothing in forensic casework. The analysis of organic FDR has concentrated on the detection of NG and 2,4-DNT by HPLC/PMDE. The HPLC/PMDE system requires a cleanup and concentration of samples containing organic residues for optimum performance.

The technique developed by Lloyd for the cleanup and concentration of organic FDR and explosive residues was assessed. The technique was found to be laborious, time-consuming, and unsuitable for the large number of samples processed by this laboratory; consequently, the extraction system has been adapted and optimized. The revised suction sampling apparatus was coupled to an automated robotic system (the Millilab 1A workstation) for the extraction and cleanup/concentration of the organic FDR using a solid phase extraction (SPE) system containing Chromosorb 104 and Amberlite XAD-4. A sensitive GC-MS method was developed for the analysis of DPA, EC, MC, camphor, and phthalates. An existing HPLC/PMDE system was adapted so that the automated deoxygenation and injection of samples for the detection of NG and 2,4-DNT were achieved. GC-MS and HPLC/PMDE analyses were performed in fractions of the same extract. The system has been applied to routine firearms casework for a trial period to assess its evidential value.

Development of the SPE System

The Millilab 1A workstation is a personal computer-controlled automated robotic system which performs sample extraction from filters and SPE devices according to user-defined programs. This was used to compare the efficiency of different SPE materials, as manual error is substantially reduced. The results are listed in Table 26.1.

Note: The relative standard deviation of the percentage recovery of the organic residues from the Chromosorb 104-Amberlite XAD-4 SPE column ranged from 3.5% for 1,3-DNB to 5.5% for DPA.

It was demonstrated that the recovery of residues from the Chromosorb 104-Amberlite XAD-4 SPE column prepared in the laboratory was more

Table 26.1 Average Recovery of 10 ng/µl Standard Containing Organic FDR by SPE on the Millilab Workstation

Organic FDR	Recovery (%) SPE Support Material		
	Chromosorb-Amberlite	C_{18}	Aminopropyl
NG	95	47	5
1,3-DNB	96	36	9
2,4-DNT	96	35	9
DPA	98	42	7
EC	95	39	2
MC	96	32	5

efficient (greater than 95%) compared to the commercial C_{18} (32% to 47%) and aminopropyl SPE columns (2% to 9%). This confirms the work of Lloyd[218] who found that Chromosorb 104 and Amberlite XAD-4 were the most efficient supports for the recovery of organic explosive residues from relatively polar solvents. To reduce the minimum volume required to elute the organic residues from the 1.5-ml SPE columns prepared in the laboratory, 40 mg of support material was used. Using these columns a 1.4-ml extract from the Deldrin unit is cleaned and concentrated to 300 µl.

The Chromosorb 104-Amberlite XAD-4 SPE column prepared in the laboratory allows full automation of the extraction process on the Millilab workstation. Subsequent experiments were performed using these SPE columns.

Efficiency of Millilab Extraction

The efficiency of extraction of organic FDR from the Deldrin unit and the subsequent SPE cleanup and concentration using the Millilab workstation were assessed. The results are listed in Table 26.2.

It was found that the recovery of organic FDR was reduced when the Deldrin filter unit was used (57% to 78% recovery compared to 95% recovery from the SPE columns). This may be explained by the presence of garment fibers and debris recovered with the FDR. The more material present within the Deldrin unit, the more difficult it is to extract the FDR with a given volume of acetonitrile (total extract 1.4 ml). Using a greater volume of acetonitrile poses problems with the subsequent 1:9 dilution of extracts from SPE. The Millilab workstation is limited to using 160 × 10 mm tubes for dilution (a total workable volume of 14 ml).

When examining "dirty" garments, a number of Deldrin units may be required to cover the entire surface as a result of the fluoropore filter

Table 26.2 Recovery of Organic FDR from Deldrin/SPE Units Using the Millilab Workstation

Organic FDR	Extraction Efficiency (%)
NG	78
1,3-DNB	72
2,4-DNT	74
DPA	57
EC	60
MC	67

Table 26.3 Average Recovery of 10 ng/µl Standard from a Swab (filter)

Organic FDR	Recovery %
NG	94
1,3-DNB	97
2,4-DNT	100
DPA	98
EC	88
MC	101

becoming clogged with material, hence reducing the suction efficiency. An attempt to use a 20-µm prefilter to prevent clogging was abandoned because this resulted in reduced recovery of inorganic FDR.

Swabs were examined to determine the efficiency of recovery of organic FDR, both without and with the SPE cleanup step. Results are presented in Table 26.3 and Table 26.4, respectively, which represent average recoveries from four tests.

This represents a very satisfactory recovery of organic FDR from swabs. The lower recovery from clothing is thought to be due to debris from the clothing inhibiting the extraction of organics. The larger volume of the Deldrin unit compared to the swab extraction apparatus, that is, different volume to solvent ratios, may be another factor. It must also be borne in mind that pure compounds may not necessarily reflect the behavior of actual organic FDR.

Table 26.4 Average Recovery of 10 ng/µl Standard from a Swab (filter and SPE)

Organic FDR	Recovery %
NG	88
1,3-DNB	82
2,4-DNT	80
DPA	93
EC	81
MC	94

Table 26.5 Analysis of FDR Recovered from Clothing (six shots)

Garment	Organic FDR (ng)				Inorganic FDR (No. of Particles)			
	NG	2,4-DNT	DPA	EC	Pb, Sb, Ba	Sb, Ba	Pb, Sb	Pb, Ba
Laboratory coat	976	39	4.6	1	3	2	71	3
Sweatshirt	1,273	39	7.4	2.2	30	5	175	13
Wool sweater	730	10	1.7	0.5	1	—	34	6

Table 26.6 Analysis of FDR Recovered from Clothing (one shot)

Garment	Organic FDR (ng)			Inorganic FDR (No. of Particles)			
	NG	DPA	EC	Pb, Sb, Ba	Sb, Ba	Pb, Sb	Pb, Ba
Laboratory coat 1	775	3.5	2.6	2	1	1	—
Laboratory coat 2	910	8.7	4.3	3	—	7	14

Recovery and Analysis of FDR from Clothing

The efficiency of the technique to recover and detect FDR from different types of clothing worn during the firing of six rounds of ammunition from a revolver was assessed. Results are given in Table 26.5.

FDR were recovered from all garments, with the sweatshirt and the laboratory coat giving better recovery than the woolen sweater. In all cases organic and inorganic FDR were easily identified. It was assumed at the start of the experiment that the woolen sweater would have the best retention of FDR but this was not reflected in the results. It is suggested that a reason for this could be the suction sampling procedure which works best on flat/tight weave garments, such as the laboratory coat and the sweatshirt.

The suction sampling and analysis techniques were repeated to determine if organic and inorganic FDR could be detected on clothing worn during the firing of one round of ammunition from a revolver. The results are listed in Table 26.6.

Detectable quantities of organic and inorganic residues were recovered from the laboratory coats. The amounts of NG, DPA, and EC detected were well above the detection limits of the systems. A smaller number of inorganic FDR particles were recovered from the laboratory coats compared to the garments doped with six shots.

Table 26.7 Survey of Clothing Submitted to the Laboratory for FDR Analysis

Case	Clothing	Organic FDR (ng)		Inorganic FDR (No. of Particles)			
		NG	2,4-DNT	Pb, Sb, Ba	Sb, Ba	Pb, Sb	Pb, Ba
F1	(a) Upper front body/cuffs	124	93	—	—	—	—
F1	(b) Pockets	300	4	—	—	1	—
F3	(a) General outer/body	124	—	—	—	—	—
F3	(b) General outer/body	2,068	4	—	—	—	—
F8	(a) Front pocket	1,685	—	—	—	1	—
F13	(a) Mask	—	—	1	1	1	1

Survey of Clothing Submitted to the Laboratory for FDR Examination

For a trial period of 3 months, clothing submitted to the laboratory for inorganic FDR analysis was also examined for organic FDR. Organic residues detected during the trial period were not used as evidence in criminal proceedings. In all 13 different firearm-related incidents (cases F1 through F13) a total of 186 exhibits were examined. One case F1 accounted for 100 exhibits. The positive results are listed in Table 26.7.

Only one exhibit, mask (a) in case F13, was positive for inorganic FDR, although no organic FDR was detected on this item. The indicative inorganic particles lead, barium were detected in 17 exhibits from four cases but, in the absence of any unique inorganic particles, they were reported as negative.

Unfortunately, it was not possible to analyze for the presence of DPA, EC, MC, camphor, and phthalates in the extracts from the 186 items as they were destroyed in a terrorist explosion at the laboratory prior to GC/MS analysis.

On the basis of this work there is poor correlation between the detection of inorganic and organic FDR in casework, and organic detection appears to be more sensitive than the detection of inorganics. However, one of the cases examined (F1) was atypical and consisted of 100 exhibits; consequently there are insufficient data to draw firm conclusions about the value of organic FDR detection.

The particle analysis method remains the preferred method for the detection of FDR and could be enhanced by the detection of organic FDR. The detection of organic FDR is a useful addition to our capability in this area as it has the potential to substantially increase the significance of particles in the "characteristic" category of the inorganic particle classification scheme for FDR identification. The ability to routinely screen all suspects, both swabs and clothing, for organic explosive residue and organic/inorganic FDR is possible and may be useful in countries with a terrorist problem. However, the means of incorporating organic FDR into the inorganic FDR system may be of interest to all involved in the detection of firearm residues.

It would be desirable to test this in casework over a period of several years for clarification in the "real-world" situation. Shortly after the development of the new system the Irish Republican Army (IRA) declared a ceasefire and the loyalist terrorist groups eventually followed suit. Consequently, the system has had very limited casework application experience. Nevertheless the results to date are very encouraging.

Current Method for Organic FDR Detection

Although automation of the extraction and SPE stages has advantages, it was unsuitable for many of our cases as it was very time-consuming (40 minutes per sample), requiring overnight running, and the Millilab could only accommodate 12 Deldrins per run. This is a disadvantage whenever there is an urgent case to be done, perhaps in the middle of the night, with the police wanting a verbal report prior to interviewing suspects. It was decided not to opt for an automatic process, but to do the sample manipulations manually.

It was also decided to use GC/TEA in preference to HPLC/PMDE for several reasons. GC/TEA is more specific for nitro groups, more suited to automation, faster, safer and simpler to use, and does not require a sample cleanup/concentration step, thereby providing better quantification as there is less sample manipulation. As a consequence a modified analytical scheme was devised, the details of which are given later in this chapter.

To evaluate the apparent greater sensitivity of organic FDR relative to inorganic FDR it was decided to conduct outdoor firing tests. For this purpose it was necessary to identify suitable ammunition as containing either single- or double-based propellant. Propellant from new boxed ammunition was analyzed. Quantitative rather than qualitative analysis was done as the information, apart from being of possible interest to the current tests, could be of use for future tests. The results are presented in Figure 26.1.

Two outdoor firing tests were conducted and the firers were examined for the presence of organic FDR. Results of the organic analysis of the first series of tests are presented in Table 26.8 and those of the second series in Table 26.9.

Elaborate precautions were taken to prevent cross contamination; yet despite this NG was detected on some of the samples originating from the discharge of single-based propellants. Obviously contamination did occur, the source of which is unknown, but it is more likely to have occurred as a consequence of being in a firing range rather than from a flaw in the procedures. These tests were conducted outdoors with a strong wind blowing, and in the first series it was also raining. The conditions were not ideal for a heavy deposition of FDR. Hands and face were sampled immediately after firing

Sample	Calibre	NG%	DNT%	DPA%	EC%	A nitro DPA	Other
?2	7.62 × 39 mm	–	–	0.8	–	Present	Camphor 1.1%
LAPUA 7.62 × 39	7.62 × 39 mm	–	–	0.7	3.3	Present	
L1 T Z A3	.223"	11.1	–	0.96	–	–	Large phthalate and glycol peak
HP 84 5.56	.223"	9.4	0.1	0.43	–	–	Large phthalate peak
PSD 88	.223"	9.6	–	0.62	–	Present	Large phthalate peak
91 RORG	.223"	6.8	–	0.01	1.3	–	MC 4.3%
IMI .223 REM	.223"	5.8	–	0.6	–	Present	Large phthalate peak
Winchester .223 REM	.223"	4.92	0.24	0.5	0.08	Present	Large phthalate peak
LAPUA .223 REM	.223"	–	–	0.45	5.7	Present	
NORMA .357 MAG	.357" Magnum	–	–	0.46	0.42	Present	Large phthalate
GECO .357 MAG	.357" Magnum	–	–	0.83	–	Present	
PMC .357 MAG	.357" Magnum	6.33	–	0.73	–	Present	Large phthalate peak
IMI .357 MAG	.357" Magnum	6.5	–	0.79	–	–	Large phthalate peak
LAPUA .357 MAG	.357" Magnum	–	0.01	0.64	2.32	Present	
Winchester 12 12 Winchester	12 Bore	9.4	0.08	0.32	–	–	Large phthalate peak
Martienoni 12 12 Genova	12 Bore	–	–	0.54	0.4	–	Large phthalate peak

Figure 26.1a Quantitative analysis of propellant.

Method for Organic Firearm Discharge Residue Detection

Sample	Calibre	NG%	DNT%	DPA%	EC%	A nitro DPA	Other
Gamebore 12 12 C.B.	12 Bore	—	0.11	0.52	—	Present	
Eley 12 12 Eley	12 Bore	—	0.06	1.27	—	—	MC 0.021%
SBP 12 12 Made in Chechoslovakia	12 Bore	—	—	1.0	1.43	—	
Cheddite 12	12 Bore	—	—	0.85	—	Present	
Rwsigeco 12 12 Rottweil	12 Bore	—	1.91	0.47	—	Present	
Express 12	12 Bore	—	0.02	0.53	—	Present	
FNM 93-1	9 mmP	18.1	0.1	0.29	—	—	Large phthalate peak
83 RG 9 mm 2z	9 mmP	14.2	—	—	—	—	Large phthalate peak
FC 9 mm LUGER	9 mmP	12.5	0.03	0.33	0.03	Trace	Small phthalate peak
GECO * 9 mm LUGER	9 mmP	—	—	0.81	—	Present	
HP 91 L7A1	9 mmP	15.7	—	0.17	0.03	—	
IMI 9 mm LUGER	9 mmP	15.0	—	0.24	—	—	Small phthalate peak
MEN 92 L10A1	9 mmP	10.4	0.05	0.26	—	—	Phthalate peak
9 mm LUGER CBG	9 mmP	12.5	0.03	0.3	—	—	
9 0 070	9 mmP	—	—	1.08	0.05	—	

Figure 26.1b (Continued)

Table 26.8 Outdoor Firings with Single- and Double-Based Propellants

Firearm	Sample	GC/TEA	GC/MS
Submachine gun		—	—
Shotgun	Swab kits	—	—
Revolver	6 shot single based	—	—
Pistol		—	—
Rifle		NG 14 ng L hand	—
Submachine gun		NG 320 ng	DPA 144 ng
Shotgun	Coats	—	—
Revolver	6 shot single based	—	EC 51 ng
Pistol		—	—
Rifle		—	—
Submachine gun		—	—
Shotgun	Swab kits	—	—
Revolver	6 shot double based	—	—
Pistol		NG 48 ng L hand	—
Rifle		—	—
Submachine gun		NG 500 ng	DPA 120 ng
Shotgun	Coats	NG 300 ng	—
Revolver	6 shot double based	—	—
Pistol		NG 240 ng	—
Rifle		NG 300 ng	DPA 66 ng

—, Nothing detected.

Table 26.9 Outdoor Firings with Single-Based Propellants

Firearm	Sample	GC/TEA	GC/MS
Swab kits			
Submachine gun	2, 6, 10 shot	—	—
Shotgun	2, 6, 10 shot	—	—
Revolver	2, 6, 10 shot	—	—
Pistol	2 shot, R hand	NG 29 ng	—
Pistol	6, 10 shot	—	—
Rifle	2, 10 shot	—	—
Rifle	6 shot, R hand	NG 200 ng	EC 3 ng
Coats			
Submachine gun	2, 6 shot	—	—
Submachine gun	10 shot	NG 680 ng	DPA 150 ng
Shotgun	2, 6, 10 shot	—	—
Revolver	2, 6, 10 shot	—	—
Pistol	2 shot	—	DPA 36 ng
Pistol	6, 10 shot	—	—
Rifle	2, 10 shot	—	—
Rifle	6 shot	—	EC 33 ng

—, Nothing detected.

and then the laboratory coat was removed and packaged; thus, sampling of the firer was prompt.

Under these firing conditions, *unique and characteristic inorganic FDR particles were readily detected on all of the samples, both swab and suction.* This result conflicts with the previous work which suggested that organics are more readily detected than inorganics (see Table 26.5 through Table 26.7). There does not appear to be good correlation between inorganic and organic FDR detection and this may be due to differing retention properties of both types of residue.

It is noteworthy that throughout the course of this work, organic residue originating from the discharge of single-based propellants has rarely been detected. EC was detected in FDR from single-based propellants in LAPUA .223" and .357" Magnum calibers (Table 26.8 and Table 26.9); however, EC occurs at a relatively high concentration (5.7% and 2.32%, respectively) in both these propellants. It was concluded that, due to the low concentration of detectable constituents in single-based propellants, it was not possible to reliably detect them in discharge residue deposited on the skin and clothing surfaces of a firer using current instrumentation and techniques. There appears to be good correlation between the detection of NG and other organic constituents of propellant, such as DPA and EC.

On the basis of this work the following system has been adopted:

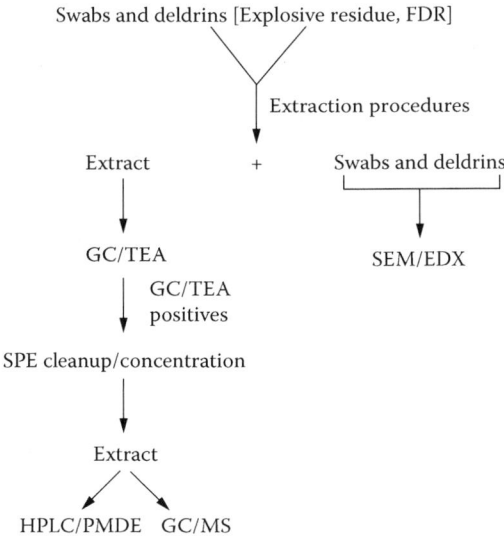

Organic FDR detection is a useful additional technique, despite the fact that single-based propellant constituents cannot be reliably detected. The current system uses GC/TEA as a rapid screening technique for NG and 2,4-DNT, and only positive samples need to be subjected to the SPE cleanup/concentration procedure. The method is flexible in that the extract can be analyzed by a range of analytical instrumentation. There does not appear to be good correlation between the detection of organic and inorganic FDR which emphasizes the need for combined instrumentation, that is, GC/TEA, GC/MS, and SEM/EDX. Details of the method devised are as follows.

Sampling of Clothing

Clothing is sampled by suction using the revised suction sampling device (see Figure 25.2). Samples taken are stored at –20°C until required. The Deldrins are reusable and are washed according to the procedure given in reference 219.

At least 10 minutes should be spent suction sampling a garment. Contamination detection measures include a dry acrilan swab from the outside of the exhibit bag prior to opening and a suction sampling device to sample the air, bench, and the sampler's hands and sleeves immediately before the examination (combined control sample). (The suction sampling apparatus can also be used for monitoring the laboratory premises for FDR/explosive residue contamination, by air sampling at different locations.) Contamination avoidance measures include hand-washing and the wearing of clean disposable laboratory coats, the use of disposable paper sheets on the workbench and the washing of the worktop with 2% v/v hydrochloric acid, followed by 2% v/v bleach, followed by acetone, both before and after an examination.

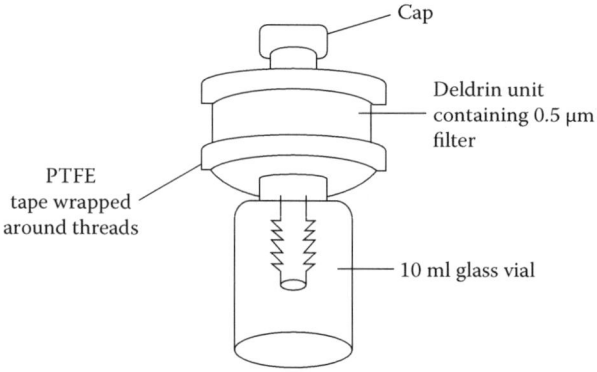

Figure 26.2 Apparatus for the extraction of organics from Deldrin.

Recovery of Organic Residue from Deldrin Unit

The extraction apparatus is prepared as illustrated in Figure 26.2, with all the components clearly labeled with the sample number.

First, 500 µl of isopropanol containing 0.6 ng/µl of 1,3-DNB (internal standard) is pipetted into the Deldrin and allowed to sit for 2 minutes. Then 500 µl of diethyl ether is pipetted into the Deldrin and allowed to sit for 5 minutes before the apparatus is centrifuged at 1,500 rpm for 2 minutes. A further 500 µl of diethyl ether is added, allowed to sit for 5 minutes, and again centrifuged at 1,500 rpm for 2 minutes. The lid is replaced on the Deldrin unit and the unit retained for inorganic FDR analysis. The organic extract is transferred to a 2 ml CV glass vial for GC/TEA analysis, after which positive samples are cleaned up and concentrated by SPE and analyzed by HPLC/PMDE for confirmation and then by GC/MS. The extracts are stored at −20°C while awaiting analysis.

Control swabs are taken from the workbench and from the operator during the organic extraction procedure for Deldrins and swabs.

Recovery of Organic Residue from Swab

The extraction apparatus is prepared as illustrated in Figure 26.3, with all the components clearly labeled with sample number.

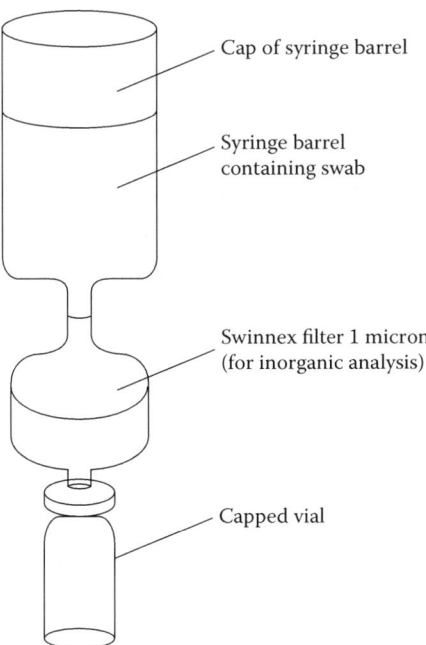

Figure 26.3 Apparatus for the extraction of organics from swab.

First, 750 µl of diethyl ether containing 0.4 ng/µl 1,3-DNB (internal standard) is pipetted into the syringe barrel containing the swab and allowed to sit for 2 minutes before centrifugation at 3,000 rpm for 1 minute. The extract is then analyzed by GC/TEA, after which positive samples are cleaned up and concentrated by SPE and analyzed by HPLC/PMDE for confirmation and then by GC/MS. The Swinnex and syringe barrel containing the swab are packaged and retained for inorganic FDR analysis. The extracts are stored at −20°C.

Note: The purpose of the in line 1-µm filter is to retain any inorganic particles removed from the swab by the organic extraction. The filter and filter holder are an integral part of a subsequent concentration/cleanup procedure for inorganic FDR as outlined in reference 220. The filter unit consists of a 13-mm-diameter Swinnex disposable filter holder containing a 13-mm-diameter, 1-µm pore size, fluoropore membrane filter (Millipore FALP 01300).

Solid Phase Extraction Procedure

Chromosorb 104, mesh size 125 to 150 µm, and Amberlite XAD-4 are the SPE materials and prior to use both materials are prepared and cleaned according to the procedure recommended in reference 221. In the procedure, 40 mg of a 3:1 mixture of Chromosorb 104:Amberlite XAD-4 is packed between frits into 1.5 ml SPE tubes as illustrated in Figure 26.4. The tube is clearly labeled with the sample number.

The SPE tubes are used in conjunction with a Visiprep D-L™ vacuum chamber complete with disposable flow control valve liners.

The SPE tubes are rinsed with 1.5 ml of acetonitrile (ACN) immediately prior to use in order to eliminate possible contaminants. The tubes are then conditioned with 1.5 ml deionized water to activate the sites in the support material. To ensure that the sites remain active sufficient water is retained to just cover the top frit. Organic extracts from swabs are blown down to near dryness (~20 µl) in an atmosphere of nitrogen, redissolved in 100 µl of ACN

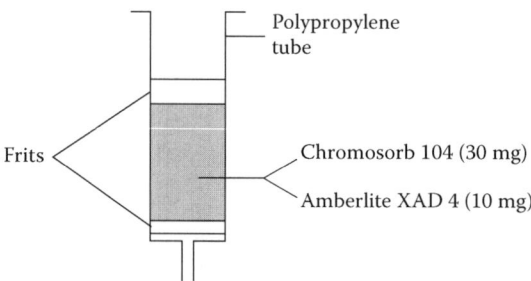

Figure 26.4 Solid phase extraction tube.

and diluted 1:10 with deionized water (deionized water reduces the affinity of organic FDR and explosive residue for ACN and IPA, permitting binding to the solid phase). The samples are then passed through the SPE tube at a rate no greater than 4 ml per minute. The SPE tube is then washed with deionized water and allowed to dry.

The internal standard and any organic FDR/explosive residue are then eluted from the SPE tube with 300 µl ACN into tapered 1.1 CTVG vials, and the extract is now ready for HPLC/PMDE and GC/MS analysis. The internal standard should be present at a concentration of approximately 1 ng/µl.

Recovery of Inorganic FDR from Deldrin

After extraction of organic FDR/explosive residue, the 0.5 µm filter is removed from the Deldrin unit and placed in a 150 ml glass beaker. The filter holder interior and cap interior are rinsed with petroleum ether (120°C to 160°C boiling range) into the same beaker and the volume made up to approximately 20 ml, after which the beaker is ultrasonicated for 20 minutes and the contents allowed to settle.

The suspension is then filtered through a concentration/cleanup system[220] consisting of two in line 13-mm-diameter filters, that is, a 25-µm wire mesh initial coarse filter and a final 1-µm fluoropore filter (Millipore FALP 01300), each of which is housed in a Swinnex filter (Millipore SX 0001300). After filtration the 1-µm filter is placed on a 13-mm-diameter aluminum SEM sample stub using double-sided adhesive tape, and carbon coated using a Biorad E6430 automatic vacuum controller after which it is analyzed by SEM/EDX for the presence of inorganic FDR.

The Deldrin filter holders and glassware are reused after cleaning using the procedure given in reference 219.

Recovery of Inorganic FDR from Swab

After extraction of organic FDR/explosive residue, the swab is placed in a glass bottle and 75 ml of dimethylformamide (DMF) added. The DMF is allowed sufficient time to dissolve the acrilan fiber and the particle suspension is filtered through the same concentration/cleanup procedure as already described, using the Swinnex holder involved in the initial organic FDR/explosives residue extraction. The 1-µm filter is then placed on a stub, carbon coated, and examined by SEM/EDX.

Contamination Avoidance

There are three distinct stages in the examination of items for FDR and explosive residue: sampling, sample preparation, and analysis. During all stages elaborate contamination avoidance procedures are employed in addition to the taking of control samples at each stage in the process. "Blanks" are included to serve as a check on the apparatus, materials, and reagents used, and the possibility of carryover from sample to sample. If contamination did occur, not only would it be identified at an early stage, but the procedure during which it occurred and the location at which it occurred would also be identified.

Statement of Witness Reports

The following style of report has been adopted for reporting cartridge discharge residue casework.

The content of the SOW report falls into four main categories following the introductory part: (1) what was done, (2) results, (3) comments, and (4) conclusion. Recommended wording of statement of witness (SOW) reports is similar to the following:

Category 1

"Items were examined for residue originating from the discharge of a cartridge, as used in firearms or blank firing devices. The identification of inorganic discharge residue particles, which originate mainly from the bullet and primer, is based on a combination of elemental composition and morphology and the particles fall into two groups, namely, those which can be conclusively identified as cartridge discharge residue and those which, while indicative of cartridge discharge, also arise from a limited number of occupational or environmental sources. The items were also examined for the presence of organic discharge residue which originates from the propellant."

Category 2

"X particles conclusively identified as cartridge discharge residue and Y particles indicative of cartridge discharge residue were detected on item _____."

"X particles indicative of cartridge discharge residue were detected on item _____, but none were found that could be conclusively identified as cartridge discharge residue."

"Nothing of significance regarding inorganic cartridge discharge residue was detected on item ____."

"The chemical compounds 'a' and 'b,' organic compounds known to be used in the manufacture of propellants, were detected on item ____."

"Nothing of significance regarding organic discharge residue was detected on item ____."

Category 3

"Deposition and distribution of cartridge discharge residue is a very random process depending on numerous factors including environmental conditions, type and condition of firearm, type of ammunition, and so forth. Persistence depends on the type of surface, activity after firing, and environmental conditions. Transfer can occur from surface to surface, and the ease of transfer will affect the persistence. Due to the factors involved, the absence of cartridge discharge residue cannot exclude someone from involvement in a firearms-related incident. The presence of cartridge discharge residues should be interpreted in conjunction with all other available information."

Category 4

"The information supplied to me in this case, along with the results, strongly supports/supports/does not support the proposition that the test subject was involved with firearms."

References

215. J. B. F. Lloyd, "High-Performance Liquid Chromatography of Organic Explosives Components with Electrochemical Detection at a Pendant Mercury Drop Electrode," *Journal of Chromatography* 257 (1983): 227.
216. J. B. F. Lloyd, "Clean-up Procedures for the Examination of Swabs for Explosives Traces by High-Performance Liquid Chromatography with Electrochemical Detection at a Pendant Mercury Drop Electrode," *Journal of Chromatography* 261 (1983): 391.
217. S. J. Speers, K. Doolan, J. McQuillan, and J. S. Wallace, "Evaluation of Improved Methods for the Recovery and Detection of Organic and Inorganic Cartridge Discharge Residues," *Journal of Chromatography* 674 (1994).
218. J. B. F. Lloyd, "Adsorption Characteristics of Organic Explosives Compounds Typically Used in Clean-up and Related Trace Analysis Techniques," *Journal of Chromatography* 328 (1985): 145.
219. J. S. Wallace, and W. J. McKeown, "Sampling Procedures for Firearms and/or Explosives Residues," *Journal of the Forensic Science Society* 30 (1993): 107.
220. J. S. Wallace, and R. H. Keeley, "A Method for Preparing Firearms Residue Samples for Scanning Electron Microscopy," *Scanning Electron Microscopy* 2 (1979).

221. J. B. F. Lloyd, "Liquid Chromatography with Electrochemical Detection of Explosives and Firearms Propellant Traces," *Analytical Proceedings* (London) 24 (1987): 239.

Conclusion

FDR has been the subject of much research and development work, over many years, in many countries, and by many workers. The following text outlines my thoughts about the nature and value of FDR work at the present time. Consider the following.

(a) Reliability of Deposition

In the process of discharging a firearm, residue may or may not be deposited on the firer and when deposited the residue varies markedly in concentration. This occurs even with repeat firings under "identical" conditions where the only obvious variables are the date and time.

Tests conducted at the NIFSL, involving both indoor and outdoor firings, under "identical" conditions revealed varying degrees of success. A few had nothing of significance detected, some had a large number of particles detected, and the majority had a moderate number of particles present. (It was noted that a .25 ACP caliber "baby" Browning pistol, firing a full magazine (six), always gave a large particle count on the firing hand.)

(b) Retention

In addition to the random nature of FDR deposition, the retention time is very low, disappearing rapidly from the hands, face, and head hair but remaining longer on the upper outer garment. The residues are chemically stable and their persistence obviously depends entirely on what happens to the clothing after the incident. If the clothing is left undisturbed, the FDR will remain indefinitely. If the clothing is worn after deposition, there will be a loss as a consequence of normal activity, namely, motion, transfer, wind, rain, and so forth. (This is less applicable to FDR inside pockets; pocket interiors are our most fruitful area.)

(c) Distribution

Excluding suicides and dead suspects, distribution (where FDR is found and its concentration in that area) is of very limited practical value and interpretation of distribution pattern needs caution. FDR on the hands can be readily transferred, in the course of normal activity, to the face, head hair, and clothing (including pocket interiors). Distribution has, in a few unusual cases, been an important factor.

(d) Interpretation

FDR on clothing could have been there before the shooting incident, or deposited after the shooting incident in the time period between the incident and apprehension of the suspect (seizure of clothing). It may not be directly attributable to the incident. A positive result for FDR on the hands, face, or head hair is strongly indicative of recent deposition. *The absence of FDR cannot exclude a suspect and the presence of FDR on a suspect does not necessarily mean that the suspect discharged a firearm.*

The suspect may have been standing close to a discharging firearm, may have picked up a recently discharged firearm, may have cleaned a firearm, may have picked up spent cartridge cases, may have had contact with the target, and so forth. However, the presence of FDR does demand an explanation from the suspect. As "FDR" could be present from nonfirearm sources, for example, cartridge-operated tools, occupational/environmental sources need to be considered.

No longer can a "unique" FDR particle be described as such as they occur from nonfirearm sources. The particle classification identification criteria have been reevaluated by other workers.[222-230] I suggest that the terms "highly indicative" and "indicative" to the discharge of a cartridge (CDR) replace the previous categories. Other identification criteria need to be developed for lead-free (Sintox) ammunition and occupational/environmental sources of strontium and titanium explored. Also the detection of such residue on the perimeter of bullet holes and on the target in close range shootings needs attention. Some workers are starting to address these aspects.

(e) Particle Composition

The deposition of FDR is a very random process and the elemental composition of discharge particles is very heterogeneous. I have read scientific papers using elemental content and inter-elemental ratios of individual particles to

Conclusion

identify primer type and even the ammunition manufacturer. In my opinion some of these claims are very ambitious as one is rarely fortunate enough to detect a sufficient number of particles in casework to enable such conclusions. Considering that lead can originate from the primer and/or the bullet; antimony can originate from the primer and/or the bullet; barium can originate from the primer or propellant; mercury could be present but not detected; tin can originate from the primer, bullet, or propellant; such conclusions, unless based on a large number of particles, are questionable. It is highly desirable to examine the inside of the spent cartridge case or cases (if present) involved in the incident before suggesting primer type.

(f) Contamination Risk

One of the problems with trace evidence is the possibility of contamination/cross contamination. Contamination has already been mentioned in chapter 24 and contamination avoidance must be a priority. If it cannot be shown that contamination is unlikely to have occurred, then the significance of laboratory positive results will be substantially diminished.

Comments

The above (a) to (f) considerations raise the question "is it worthwhile examining suspects for FDR?" This is a very difficult question to answer but it would be an interesting topic for a debate among police, forensic scientists, those responsible for financing the work, and the legal profession.

FDR on a suspect is "supporting" evidence, although it has been an important factor in several cases. FDR on the bullet surface and on the target is both useful and reliable when identifying bullet holes, differentiating between entry and exit holes, estimating range of fire, and identifying bullet type (for example, tracer, incendiary, and so forth).

The search for a simple, reliable, fast, inexpensive, and conclusive test for FDR has yet to be realized. There is no doubt that the particles analysis method is the most desirable and informative method to date, but it is more suited to a moderate FDR caseload. In Northern Ireland we had a heavy FDR caseload and multisuspect cases were common. In one incident we had 38 suspects, with three or four upper outer garments each, to be examined. For a heavy FDR caseload the fast, inexpensive FAAS bulk elemental system is attractive, particularly considering the downgrading of the "uniqueness" of the particles detected coupled with the lengthy and costly procedures of the particle analysis method. The properties of FDR plus the lengthy and costly procedures of the particle analysis method, the limitations of a positive result and the facts of the incident should be discussed by an experienced forensic scientist and the police officer in charge of the case, to determine whether or not to proceed. An early consultation should at least identify the items/areas to be sampled with the greatest likelihood of success. The argument that all crimes involving the use of firearms are serious and demand the fullest investigative effort is difficult to fault. However, it must be tempered with realism and common sense.

(It is interesting to note that the automobile features often in the interpretation of FDR results. In the Price method[231] the hands of suspects were examined for the presence of lead in particulate form. This method was discredited when it was discovered that automobile exhausts eject numerous lead particles into the environment from leaded petrol. When bulk elemental analysis methods were developed it was shown that lead could originate from leaded petrol and the car battery terminals. Antimony was also present in battery terminals and barium was present in motor oil. Finally, in the particle analysis method it has been claimed that lead, antimony, and barium "unique" FDR particles can originate from automobile brake linings.)

The Future of the Firearm

What is the future of the firearm? Caseless ammunition and suitable firearms are currently available in 5.7 mm UCC and 6.0 mm UCC calibers, with the ammunition activated electrically. The ammunition is considerably smaller and weighs approximately one third of conventional ammunition, both of which are advantageous, particularly in a war situation, as more ammunition could be carried by an individual. The forensic implications of such a firearm/ammunition combination are that there would be no spent cartridge cases for comparison purposes. Electrical ignition of conventional ammunition is currently being researched and if successful would do away with the primer and reduce the weight of a round of ammunition. Implications for forensic science examination would be no firing pin impression on the cartridge case for comparison purposes.

Directed-energy weapons are currently being researched and developed. Instead of transferring energy via a projectile (kinetic energy), they transfer energy to the target by some other means, for example, they cause physical damage to the target with particle or electromagnetic beams (lasers, masers). All such weapons require high electric power and are currently not a practical proposition for handheld weapons.

With rapid advances in technology there is no doubt that firearms and ammunition will continue to develop and it is not beyond the realm of possibility that firearms could be replaced by some other form of handheld weapon.

References

222. J. S. Wallace, and J. McQuillan, "Discharge Residues from Cartridge-Operated Industrial Tools," *Journal of the Forensic Science Society* 24 (1984): 495.
223. F. S. Romolo, "Identification of Gunshot Residue: A Critical Revue," *Forensic Science International* 119, no. 2 (2001): 195–211.
224. B. Cardinetti, C. Ciampini, C. D'Onofrio, G. Orlando, L. Gravina, F. Ferrari, T. D. Di, and L. Torresi, "X-ray Mapping Technique: A Preliminary Study in Discriminating Gunshot Residue Particles from Aggregates of Environmental Occupational Origin," *Forensic Science International* 143, no. 1 (2004): 1–19.
225. P. Bergman, E. Springer, and N. Levin, "Hand Grenades and Primer Discharge Residues," *Journal of Forensic Sciences (JFSCA)* 36, no. 4 (1991): 1044–52.
226. A. Zeichner, and N. Levin, "More on the Uniqueness of Gunshot Residue (GSR) Particles," *Journal of Forensic Sciences* 42, no. 6 (1997): 1027–8.
227. L. Garofano, M. Capra, F. Ferrari, G. P. Bizzaro, D. Di Tullio, M. Dell'Olio, and A. Ghitti, "Gunshot Residue: Further Studies on Particles of Environmental and Occupational Origin," *Forensic Science International* 103, no. 1 (1999): 1–21.
228. C. Torre, G. Mattutino, V. Vasino, and C. Robino, "Brake Linings: A Source of Non-GSR Particles Containing Lead, Barium, and Antimony," *Journal of Forensic Sciences* 47, no. 3 (2002): 494–504.
229. P. V. Mosher, M. J. McVicar, E. D. Randall, and E. H. Sild, "Gunshot Residue—Similar Particles Produced by Fireworks," *Canadian Society of Forensic Science Journal* 31, no. 2 (1998): 157–168.
230. J. R. Giacalone, "Continuing the Quest for Non-Firearm Sources of Gunshot Residue," *Scanning* 25, no. 2 (2003): 69.
231. G. Price, "Firearms Discharge Residues on Hands," *Journal Forensic Science Society* 5 (1965): 199.

Index

2,4-dinitrotoluene, 63
2,6-dinitrotoluene, 63
.357" Magnum caliber ammunition, 224
35-NF primer composition, 44
.38" special ammunition, 224

A

ABC bullet, 72
Acetonitrile (ACN), 266
Aconitin, bullets poisoned with, 84–85
Adhesive lifts, hand sampling via, 132
Aircraft load, special caliber, 84
Aluminum alloys, in firearm construction, 97
Aluminum bullets, 83
Ammonium nitrate, 60
Ammunition
    chemical aspects, 33
    current trends toward lighter, 93
    defined, 9
    historical aspects, xi, 11, 23–27
    typical types, 9
Ammunition analysis, 183
    interpretation of, 188–199, 200–201, 203
    miscellaneous ammunition components, 188
    primer types, 183
    propellant analysis, 183–188
Ammunition components, analysis of miscellaneous, 188, 195–200
Anthrax, bullets poisoned with, 84
Anticorrosion measures, 91
Antimony
    as bullet hardening agent, 70, 189
    emission upon discharge, 55
    in FDR particles, 139
    forensic detection, 108
    health hazards, 223
    in plated shot, 76

    quantitative detection of, 108
    use in hardening bullet lead, 20
Antimony-free primers, 179
    discharge particles from ammunition with, 180
Antimony sulfide, 166
    in priming compositions, 41
Arcane bullet, 72
Armor-piercing bullets, 71–72
Artificial rusting, 98
Artillery, 4
Atomic absorption spectrophotometry, 109–110
Average recovery
    organic FDR from Deldrin/SPE units, 255
    organic FDR on Millilab workstation, 254
    from swab filter, 256
    from swab filter and SPE, 256

B

Back flushing, 253
Back thrust, 37
Backscattering, 112
Bacon, Robert, 13, 14
Ballistics, defined, 4
Bang sticks, 96
Barium
    emission on discharge, 55
    in FDR particles, 139, 179
    in flare residues, 150
    health hazards, 223
    as indicative *vs.* unique particles, 143
    quantitative detection of, 108
    residue from barium-free primers, 173
Barium nitrate, 53
    as potassium chlorate replacement, 45
Barnes soft point, 70

Barnes super solid, 70
Baton rounds, 87
    analysis, 179–181
Berdan primer design, 39, 40
Binding agents
    gum arabic, 41
    in priming compositions, 42
Bingham Limited, 80
Bismuth compounds, in propellants, 62
Bismuth shot, 76
Black powder, 13
    analysis of unburned, 167
    diminished use of, 59
    fouling due to, 43
    postcombustion residue, 59
Black powder ammunition, discharge residue from, 166–169, 168
Black sealant, 181
Blank ammunition, 95–96
    chemical aspects, 95–96
Blank cartridges, 95
    homogeneous character of discharge gases, 149
    limited range of discharge particle types, 147
    particle classification scheme, 143–149
    residue in, 144
    similarity in composition to live ammunition, 145
Blank-firing imitation/replica firearms, 95
Blitz-Action-Trauma (BAT) ammunition, 83–84
Blow back principle, 216
Bluing process, 98
    formulae for, 99
Bogus head stamps, 3i
Borcharott, Hugo, 31
Bore
    defined, 8
    leading of, 20
Boxer primer design, 39, 40
Brass
    and ammunition velocity, 37
    as most popular cartridge case/primer cup material, 188
    in primer cups, 39
    specifications, 36
    use in cartridge cases, 35
Brass cartridge cases, 43
Brass-jacketed bullets, particles on new ammunition, 158
Breech residue, 214

sampling box for, 215
Brown powder, 60
Browning process, 98
    formulae for, 99
Bulk elemental analysis methods, lack of specificity, 110
Bullet base designs, 68
Bullet core designs, 67–68
    materials, 70
Bullet deformation, 20
Bullet fragmentation, 163
Bullet fragments, chemical comparison, 103
Bullet hole perimeters, 172–173
    elemental levels, 176–177
    mercury deposition on, 208, 218
    test ammunition, 174–175
Bullet jacket designs, 70
Bullet jacket material, determining from bullet hole perimeter residue, 175
Bullet jackets, 20, 181
    composition of, 20–21
    materials, 69
    residue particles originating from, 155
Bullet lubrication, 20
Bullet particles, 124
Bullet size, 19
Bullet tips, 71
Bullet velocity
    and jacketing, 68
    and number of discharge residue particles, 124
Bullet weight loss, on firing, 157, 160–161
Bullet wipe patterns
    and firing angle, 172
    mercury in, 212
Bullets, 67–71
    armor-piercing, 71–72
    cylindro-conoidal, 20
    cylindro-ogival, 20
    elongated, 19
    filled, 71
    hexagonal, 20
    historical aspects, 19–21
    incendiary, 74–75
    mercury remaining post-discharge, 212
    misuse of term, 9
    revolver and pistol metal-piercing, 72
    as source of FDR particulates, 123
    tracer bullets, 72–73
Burning rate
    and granulation size, 59
    and physical propellant shape, 58

Index

C

Calcium, as indicative *vs.* unique particles, 143
Calcium carbonate, as propellant neutralizer, 61
Caliber, basis of, 8
Cannonlocks, 29
Cannons, in firearms history, 29
Caps, 39–40
Carbon steels, in firearm construction, 97
Cartridge cases, 181
  base, 37
  brass, 188
  chemical aspects, 35–38
  high-velocity, 37
  residue in spent, 144, 145
  sampling experimentation, 141–142
Cartridge-operated industrial tools, 143
Cartridges
  completely combustible, 27
  current developments, 27
  defined, 9
  self-contained, 24
Case hardness, 37
Case thickness, 37
Caseless ammunition, 27, 93–94, 274
  chemical aspects, 93–94
Casework-related tests, 157
  antimony-free primers, 179
  baton round analysis, 179–181
  bullet fragmentation, 163
  bullet hole perimeters, 172–175
  bullet weight loss on firing, 157, 160–161
  discharge residue from black powder ammunition, 166–169
  effects of water on FDR, 161–162
  firearm coatings, 169
  homogeneity of propellants, 169–171
  mercury-containing particles in, 208
  particles containing mercury, 209–210
  particles from handling ammunition, 157
  persistence, 175–179
  RPG7 rocket launcher, 163–166
Centerfire cartridges, 26
  primers used in, 39
Centerfire round, 9
Chamber pressures, cartridge cases, 36
Characteristic X-ray emission, 112
Charcoal, 14
  in black powder, 59
  in gunpowder, 13
Chemical aspects
  blank ammunition, 95–96
  cartridge cases, 35–38
  caseless ammunition, 93–94
  complementary ammunition components, 91
  firearm construction materials, 97–98
  Primer caps, 39–40
  priming compositions, 41–56
  projectiles, 67
  propellants, 57–66
Chilled shot, 75
Chromatography techniques, 248
  use in FDR, 114
Chromium-molybdenum steel, in firearm construction, 97
Chromosorb 104, 266
Clean rooms, 235
Climatic conditions, and rate of loss of mercury, 220
Close range shooting
  with black powder ammunition, 167–168
  lead as indicator of, 173
Clothing
  analysis of FDR recovered from, 257
  examination of, 248–249
  FDR recovered from, 247
  membrane pore size experiment on, 245, 246
  persistence of residue on, 177
  pocket interior sampling, 178
  recovery and analysis of FDR from, 257
  sampling for FDR, 241–242, 242–246
  survey of submitted, 258–259
Clothing sampling method, 242–246
Clusters, in FDR, 125
Coated iron jackets, 189
Cobalt, detection in Nyclad bullets, 223
Cocoa powder, 60
Color coding, 91
Colored taggants, in propellants, 61
Colt, Samuel, 30
Column efficiency, in FDR techniques, 115
Combined gas chromatography-mass spectrometry (GC-MS), 117
Combustible paper cartridges, 24
Combustion products, 59
  solid and gaseous, 60
Communist Bloc countries, use of mercury fulminate primers, 200

Complementary ammunition components, 91
   chemical aspects, 91
Complex hypophosphite sales, in priming compositions, 48
Contamination
   assumptions in sampling procedures, 249
   avoidance procedures, 234–235, 264, 268
   avoiding in FDR sampling, 131, 139
   risk of, 273
Contamination avoidance kit, 236
Contamination detection measures, 264
Cooper, Joseph Rock, 25
Copper
   in blank cartridge particulates, 148
   in FDR particulates, 124
   residues from unjacketed bullets, 148
Copper alloy bullet jackets, 189
Core-Bond, 69
Corned gunpowder, 13
Crime
   experimental work review, 138
   homemade firearm construction materials, 98
   most commonly used firearms in, 5
Cross-contamination, 234–235, 273
   preventing, 250
   prevention in organic FDR detection, 259
Current force/direction, and hand deposits, 127
Cylindro-conoidal bullets, 20
Cylindro-ogival bullets, 20

D

Damp clothing, low retrieval rate of FDR, 161
Dardick Trounds, 87
Dark ignition tracer bullets, 73
Day tracer bullets, 73
Decoppering additives, 62
Definitions
   ammunition, 9
   discharge of firearm, 9–10
   firearms, 3–8
   weapons, 3
Deldrin filter, 250
   apparatus for extraction of organics from, 264
   recovery of inorganic FDR from, 267
   recovery of organic residue from, 265
   reduced recovery of organic FDR with, 255
Deposition, reliability of, 271
Dermal nitrate test, 107
Devil bombers, 149
Diazonitrophenol, as lead styphnate substitute, 47
Dibutyl phthalate, 63
Diethyl phthalate, 63
Dim ignition tracer bullets, 73
Dimethyl phthalate, 63
Dinitrotoluene, 62
Diphenylamine, 61, 62
   in caseless ammunition, 93
   as most common stabilizer in single-based propellants, 190
Diphenylamine test, 107
Directed-energy weapons, 274
Discarding sabot, 72, 81
Discharge of firearm
   defined, 9–10
   differences between firearms and blank firers, 149
   undesirable health consequences, 55
Discharge primer residue, 171
Discharge residue
   loss with time, 234
   in mercury-containing ammunition, 216
Discharged warhead, residue in, 164
Distribution, of FDR, 272
Double-based propellants, 62, 65
   analysis of, 194
   outdoor firings with, 262
Dreyse, Johann Nikolas, 24
Dreyse cartridges, 25
Dual core bullets, 70
Duration of exposure, 128
Dynamit Nobel AG, 55, 228
   development of Sintox, 224

E

Edwards E2M6 vacuum pump, 242
Electrical ignition, 274
Electroless nickel coating, 100

Index 281

Electron impact quadruple mass spectrometer, 116
Electron spin resonance spectrometry, 109, 114
Electroplated jackets, 68
Elemental composition
　on bullet hole perimeters, 176–177
　heterogeneity of, 272
　of match residues, 150
　per particle type, 148
Eleyprime, 55
Elongated bullets, 19
Environmentally friendly ammunition, 223, 224
Equaloy, 84
Ethyl centralite, 62, 191
Ethyl centralite (EC), as most common stabilizer in double-based propellants, 190
Ethyl entralite, 62
Evidence protection kit, 235–236
Experimental data, 135
　ammunition analysis, 183
　casework-related tests, 157
　instrumentation, 139–141
　lead-free ammunition, 135
　mercury-containing ammunition, 205
　objectives, sampling procedures, instrumentation, conditions, 137–142
　sampling, 141–142
Exploding bullets, 78–80, 79
Explosive residue, 142
　detection techniques, 253
Explosive/tracer rifle bullets, 73
Explosives, in priming compositions, 42
Exterior ballistics, defined, 4
Extra hard shot, 75

F

Face residue, time of deposit, 179
FDR kit design, 131
Filled bullets, 71
Film lifts, hand sampling via, 132
Firearm coatings, 169, 170–171
Firearm construction materials, 97–98
　chemical aspects, 97–98
　surface coatings, 98–100
Firearm contamination, 181

Firearm discharge residue detection techniques, 106
　detection of organics, 114–117
　flameless atomic absorption spectrophotometry, 109–110
　gas chromatography/mass spectrometry, 115–117
　Harrison and Gilroy method, 108
　high-performance liquid chromatography, 114–115
　to nanogram/picogram level, 115
　neutron activation analysis, 108–109
　paraffin test, 106–107
　particle analysis method, 111–114
Firearm discharge residue (FDR), 101, 103–106, 142
　from black powder ammunition, 166–169
　confusion of particles from blank cartridges with, 146–147
　contamination risk, 273
　detection techniques, 103–117
　development of detection techniques, 106–114
　in discharged warhead, 165
　distinct differences from starting pistol discharge residue particles, 147
　distribution issues, 272
　effects of water on, 161–162
　experimental work in, 137
　and future of firearms, 274
　heterogeneous nature of, 151, 155
　interpretation of findings, 272
　organic components, 239
　particle composition, 272–273
　particle types from promptly collected, 146
　persistence of, 178
　properties, 123–132
　random variation in concentration, 271
　recovery of organic, from clothing, 244, 247, 257
　reliability of deposition, 271
　retention, 271
　role of automobile in interpretation of, 274
　from RPG7 rocket launcher, 164
　sampling kits for, 247–248
　sampling of clothing for, 242–246
　sampling of skin and clothing for, 241–242
　as supporting evidence, 273

vs. flare residue, 150
Firearm discharge residue properties, 123
   composition, 126–127
   deposition, 127–128
   distribution patterns, 129–131
   formation, 123–125
   morphology, 125
   particle population, one shot, same caliber, 130
   persistence, 131
   quantity and composition, 128–129
   sample collection for, 131–132
   size, 125–126
Firearm discharge residue sampling, 233
   comment, 236–237
   contamination avoidance, 234–235
   evidence protection kit, 235–236
   time delays in, 233–234
Firearms
   chemical aspects, 33
   defined, 3–8, 4
   features, 5
   historical aspects, xi, 11, 29–31
   tracing to foreign suppliers, xv
   uses, 5
Firearms examination, in terrorist *vs.* non-terrorist environment, xv
Fireworks, particle classification scheme, 151
Firing angle, and bullet wipe patterns, 172
Firing location, 128
Firing mechanism, 9
Firing pin, 9
Flameless atomic absorption spectrophotometry, 109–110, 110, 141
Flare loads, 80
Flares, particle classification scheme, 150
Flash pan, 15
   introduction of waterproof, 16
Flash reducers, 62
Flechette cartridges, 87
Flintlock, 16, 30
Fluoropore filter, 255
Forensic firearms casework examination, xiii, 103
Forsyth, Alexander John, 15, 16, 42
Fouling, due to black powder residue, 43
Frangible bullets, 87
Frankford Arsenal, 44
   P-4 primer, 48
French Silver Plus, 7

Friction, and bullet weight loss, 160
Frictionators, in priming compositions, 42
Frizzen, 16
Fuels, in priming compositions, 42
Full metal jacketed bullets. *See also* Total metal jacketed bullets (TMJ)
   weight loss on firing, 157

G

Gas chromatography, 114, 115–117, 140–141, 141
Gaseous combustion products, 60
GC/TEA detection, 259
Geco lead-free ammunition, 223
   primer analysis, 224
Glaser safety slug, 82
Glassless rimfire primers, 49
Gonzales test, 107
Grains, 58
Granulation size, and burning rate, 59
Granules, 58
Grips, construction materials, 98
Gum arabic, as binding agent, 41
Gun, correct usage of term, 3
Gun cleanliness, and hand deposition, 127
Gun condition, and hand deposition, 127
Gun type, and hand deposition, 127
Gunpowder
   historical aspects, 13–14
   vs. propellants, 58
Gunshot residue (GSR), 103. *See also* Firearm discharge residue (FDR)

H

Hammer, 9
Hand cannons, 29
Hand deposits, 127, 129
   persistence of, 131
Hand loading, 77–78
Hand residue
   ease of removal, 177
   time of deposit, 179
   and time of firing, 177
Hand sampling methods, 132
Handguns, 5
Handheld weapons, 3

Hard shot, 75
Harrison and Gilroy method, 108
Head hair residue
  removal by suction sampling, 242
  time of deposit, 179
Head stamps
  on blank ammunition, 95
  bogus, 38
  details in, 37
  political misuses of, 38
Health hazards, 223
Heavy metals, in FDR particles, 112
Heavy weaponry, xiv
Heckler & Koch, caseless ammunition by, 93, 94
Hexagonal bullets, 20
High-performance liquid chromatography, 114–115, 141
Hijackers
  bullets for dealing with, 83
  special caliber aircraft load for, 84
Historical aspects
  ammunition, 11
  bullets, 19–21
  firearms, 29–31
  gunpowder, 13–14
  ignition systems, 15–17
HKGH rifle, 93
Homemade bullets, 78
Homemade firearms
  construction materials, 98
  use of household paints on, 169
Hornady FMJ, 70
Hostage takers, special bullets for, 83
Hot salt, black oxide process, 99
HPLC/PMDE, 267
  detection of NG by, 253
Hydra-Shok, 82

I

Ignition, dependence on weather conditions, 16
Ignition systems, historical aspects, 15–17
Ignition temperature, of caseless ammunition, 93
Incendiary bullets, 74–75, 79
  forensic determination of, 106
Incident report forms, 250
Indicative particles, 143, 272
  classification of, 157
  on clothing, 258
  in particle classification scheme, 154
  percentage occurrence of accompanying elements in, 152–153
  proportion of, 148, 149
Indoor fireworks, 151
Infrared spectroscopy, 114
Initial detonating agents, 41
Inorganic discharge residue, 104, 125. *See also* Firearm discharge residue (FDR)
  recovery from Deldrin, 263
  recovery from swab, 267
Instant ignition, 16
Instrumentation
  experimental work, 139
  flameless atomic absorption spectrophotometry, 141
  gas chromatography, 140–141, 141
  high-performance liquid chromatography, 141
  mass spectrometry, 141
  pendant mercury drop electrode, 141
  scanning electron microscopy, 139–140
  thermal energy analyzer, 140–141
Interior ballistics, 104
  defined, 4
Irish Republican Army (IRA), 259
Irregular particles, in FDR, 125

J

Jacket hardness/thickness, 69
Jacketed bullets, 67–68
  electroplated, 68
  particles on new ammunition, 160
  residue particulates from, 123
Jacketed hollow point (JHP) ammunition, 129

K

Kernels, 58
Kindling temperature, 59
Kneecapping incidents, 200
Kraft paper, in caseless ammunition, 93

L

Lacquer, 181
Laminated wooden stocks, 97
Land, 8
Lead
    from automotive sources, 274
    distribution between layers, 162
    effect of water on FDR levels, 162
    emission upon discharge, 55
    in FDR particles, 139
    health hazards of airborne, 223
    as indicator of bullet damage, 173
    as indicator of close range shootings, 173
    origins in FDR, 273
    proportion of particulates as, 123
    sodium rhodizonate test for, 172
Lead azide, in priming mixtures, 47
Lead bullets, 19
Lead-free ammunition, 223
    analysis, 225–227
    discharge particles from, 224, 228
    identification criteria for, 272
Lead-free primers, 55
Lead monoxide, 55
Lead nitroaminotetrazole-lead styphnate, 49
Lead-only particles
    in FDR *vs.* blank cartridge residues, 147
    in unjacketed bullet residues, 148
Lead styphnate, 53
    diazonitrophenol as substitute, 47
    as mercury fulminate replacement, 46
    in Sinoxyd-type primers, 47
    substitutes for, 49
Lead styphnate hydrate, 48
Lifting efficiency, 132
    in FDR sampling, 131
Link reports, xiv
Liquid traps, mercury recovery from, 213, 217
Lloyd method, 253, 254
Loading time, attempts to reduce, 23
Lock, stock and barrel, 10
Lubaloy, 69
Lubricants, 91, 100
Luger, George, 31

M

Machine guns, 5
    development of fully automatic, 31
Magnesium, as incendiary agent, 74
MagSafe bullets, 84
Major level, 139
Markov, George, 86
Mass spectrometry, 114, 115–117, 116, 141
    disadvantages for FDR work, 116
Matches, particle classification scheme, 50
Matchlock, 15, 30
Matrix compatibility, in FDR sampling, 131
Maxim, Sir Hiram, 31
Membrane pore size experiments, 245, 246
Mercuric and corrosive primers, 46, 55, 181, 201
Mercuric and noncorrosive primers, 55, 201
Mercuric but noncorrosive primers, 46
Mercury
    amalgamation with zinc, 205
    ammunition containing, 205 (*See also* Mercury-containing ammunition)
    amount deposited on firer, 219
    as bullet hardening agent, 70
    in bullet hole perimeter, 218
    on bullet hole perimeter, 175
    in bullet wipe, 212
    casework particles containing, 209–210
    concentration effect in breech emission, 217
    deposition on firer, 216
    disappointing percentage recovery, 215
    distribution after discharge, 208–220, 215
    distribution between particulate and vapor, 214
    end of U.S. use of, 201
    indetectability of, 207
    loss by adsorption/absorption, 218
    loss from spent cartridge cases, 220
    loss with time, 220
    low concentration on filter, 216
    origins in FDR, 273
    particles from blanks, 147
    percentage indetectable by SEM, 217
    presence in vapor/submicron particle form, 217
    prevalence of large particles, 216, 218
    recovery from liquid traps, 213

Index

remaining in gun, 212
remaining in spent cartridge case, 211
remaining on bullet, 212, 218
residue from muzzle, 216
scarcity of discharge particles containing, 138
slow rate of loss, 220
volatility of, 205
Mercury-containing ammunition, 205
amount deposited on firer, 219
distribution of mercury after discharge, 208–220, 215
experimental errors, 218
frequency of occurrence, 205–208
improved residue sampling apparatus, 219
and loss of mercury with time, 220
mercury-containing particles from, 207
mercury in bullet wipe, 212
mercury levels from primer discharge, 213
mercury levels in muzzle discharge residue, 214
mercury loss from spent cartridge cases, 220
mercury remaining in gun, 212
mercury remaining in spent cartridge case, 211
mercury remaining on bullet, 212
occurrence of mercury particles, 206
primer discharge sampling system, 213
quantitative comparison, filter and liquid traps, 217
sampling box for breech residue, 215
total mercury content, 211
Mercury-containing particles, 207
in casework, 208
Mercury fulminate, 16, 25, 30
as basis of priming compositions, 42
lead styphnate as replacement for, 46
in Northern Ireland, 205
in priming compositions, 43
use in Eastern Bloc countries, 53, 200
Mercury particles, occurrence of, 206
Mercury vapor, 219
use of acidified potassium permanganate with, 219
Metal, discovery of, 3
Metallic cartridge cases, 25
Metallic cartridges, 23
Metallic fulminates, 16
Metford, William Ellis, 20

Methyl centralite, 62, 192
Microchemical crystal tests, 114
Military training, blank ammunition for, 95, 96
Millilab workstation, 254
average recovery of organic FDR, 254
efficiency of extraction, 255–257
recovery of organic FDR from Deldrin via, 255
Minor level, 139
Misch metal, 50
Molecular luminescence, 109, 114
Multiple loads, 81–82
Multipurpose filled bullets, 75
Multisuspect cases, 235
detection of organics in, 241
in Northern Ireland, 273
Muzzle blast, 96, 105
Muzzle blast residue, 105
Muzzle discharge residue, mercury levels in, 214
Muzzle discharge sampling system, 214
Muzzle flash, 4, 105
minimizing, 61
Muzzle flash suppressors, 62
Muzzle loading, 23
Muzzle residue
membrane pore size experiment on, 245
mercury presence on, 217
Muzzle velocities, 10

N

National bullet, 72
Neutron activation analysis, 108–109
Nickel, on bullet hole perimeter, 173
Night tracer bullets, 73
Nitrocellulose, 60
in caseless ammunition, 93
stabilizers for, 61
Nitroguanidine, 62
Nobaloy, 69
Noncorrosive, nonmercuric (NCNM) priming compositions, 44, 47
U.K. shift to, 47
Noncorrosive primers
German advances in, 45
from RWS, 45
Swiss Army type, 45
U.S. military conversion to, 47

Nonmercuric and corrosive primers, 55, 201
Nonmercuric and noncorrosive primers, 55, 201
Nonmercuric but corrosive primers, 6
Nonmercuric primers, 44
Nontoxic primers, 201
Nonuniformities, in propellants, 58
Northern Ireland, xi
   background to terrorist situation, xii–xiii
   decreasing percentage of casework with positive FDR, 175
   FDR detection experimental work, 137
   indoor *vs.* outdoor fireworks, 151
   limited use of flares in, 150
   loyalist paramilitary groups, 203
   multisuspect cases in, 273
   risk of cross-contamination, 234
   RPG7 rocket launcher use, 163
   shooting incidents based on single-based propellants, 241
   suspect hand protection kits, 236
Northern Ireland Forensic Science Laboratory (NIFSL), xi, 223
Nosler partition, 70
Nuclear magnetic resonance/polarography, 114
Nyclad ammunition, 223

O

Obturation, 10
   prevention of rearward escape of gas through, 36
Organic components
   detection in FDR, 138, 242
   development of detection methods, 239
   of FDR, 239
   FDR techniques, 114–117
   recovery from clothing, 244
   sensitivity of detection, 258
   skin and clothing sampling, 241–242
   speed of detection for multisuspect cases, 241
Organic FDR detection
   apparatus for extraction from Deldrin, 264
   apparatus for extraction from swab, 265
   average recovery on Millilab workstation, 254
   contamination avoidance in, 268
   current method, 259–266
   development of method for, 253–254
   efficiency of Millilab extraction, 255–257
   outdoor firings with double-based propellants, 262
   outdoor firings with single-based propellants, 262
   quantitative analysis of propellant, 260–261
   recovery from clothing, 252
   recovery from Deldrin unit, 265
   recovery from swab, 265–266
   solid phase extraction procedure, 266–267
   by SPE system, 254–255
   statement of witness reports, 268–269
   survey of clothing submitted for, 258–259
Outdoor fireworks, 151
Outdoor firings, with single- and double-based propellants, 262
Oxidizers, in priming compositions, 42

P

Paper cartridges, 23
Paraffin test, 106–107
Parkerizing finish, 100
Partially jacketed bullets, 68
Particle analysis method, 106, 111–114, 114, 123, 129, 241, 248, 258
   for lead-free ammunition, 225–227
Particle classification scheme, 137, 143
   accompanying elements, 151
   blank cartridges, 143–149
   failure to include mercury fulminate-primed ammunition, 205
   fireworks, 151
   flares, 150
   improving, 138
   limitations and revisions, 154–155
   matches, 50
   need for revision, 143
   toy caps, 149–150
Particle composition, 126–127, 272–273
Particle population, one shot, same caliber, 130
Particles

Index

from ammunition with antimony-free primers, 180
from black powder ammunition, 168
chemical stability on clothing, 177
from handling ammunition, 157
from lead-free ammunition, 224, 228
mercury-containing, 207
on new ammunition, 158–160
on unloaded ammunition, 160
Particulates
  elemental level by particle type, 148
  for FDR, 123
Pasteboard cartridge cases, 25
Pauly, Jean Samuel, 24, 26
Pauly cartridges, 24
Pendant mercury drop electrode, 141
Percussion cap, 16, 17
Percussion primer, development of, 15
Percussion primer compositions, 42
Percussion system of ignition, 16
Percussionlock, 30
Performance, of cartridge cases, 36
Perimeter residue, 106
Persistence, 175–179
Peters Rustless primer, 46
PETN, 49, 80
Phosphatizing finish, 100
Phosphorus, as incendiary agent, 74
Piercing/incendiary/tracer bullets, 71
Pinfire cartridges, 25, 26
Pistols, 6
  defined, 7
  primer cups for, 40
  use in crime, 5
Plastic, use in shotgun cartridges, 35
Plastic bags, disadvantages in evidence protection, 235
Plasticized cellulose nitrate (NC), 60
Plasticizers
  fuel type, 60
  high energy oxidizing, 60
  methyl centralite, 192
  organic crystalline chemicals, 61
Plated shot, 76
PMC Ultra-Mag ammunition, 82
Pocket interiors, 249
  as most fruitful retention area, 271
  sampling of, 178
Point blank range, 105
Poisoned arrows, 85
Poisoned bullets, 84–86
  German designed, 85, 86

Soviet designed, 85
Polymers, in firearm construction, 97
Potassium, in FDR particles, 166
Potassium chlorate
  barium nitrate as replacement for, 45
  in priming compositions, 43
  rusting due to, 43
Potassium nitrate
  as flash suppressor, 62
  WWI shortage of, 60
Potassium sulfate, as flash suppressant, 62
Power heads, 96
Price method, 274
Primer caps, chemical aspects, 39–40
Primer composition, 40
Primer cups, 10, 39–40, 173, 181
  brass in, 188
  mercury remaining on, 213
Primer discharge
  mercury levels from, 213, 217
  muzzle discharge sampling system, 214
  sampling system, 213
Primer particles, 124
Primer type
  ammunition analysis, 183
  forensic studies of, 106
  methods for determining, 200
Primers, 9, 181
  antimony-free, 179
  categorization of, 201
  elemental content, 144
  as source of FDR particulates, 123
Priming compositions, 41–42
  binders in, 42
  characteristics of ideal, 41
  chemical aspects, 41–56
  common U.K. type, 54–55
  common U.S. types, 53–54
  complex hypophosphate salts in, 48
  double salts in, 48
  duration of flame, 42
  Eleyprime, 55
  experimental patents, 51–53
  explosives in, 42
  frictionators in, 42
  fuels in, 42
  German advances in noncorrosive, 45
  lead azide in, 47
  lead-free, 55
  misch metal, 50
  noncorrosive, nonmercuric (NCNM), 44
  nonmercuric, 44

nontoxic Sintox type, 55
oxidizers in, 42
percussion type, 42
pyrophoric metal alloys in, 50
rate of burning, 42
role in ignition, 41
sensitivity of, 41
sensitizers in, 42
six categories of contemporary, 55
Staba types, 50
stabanate, 49
stabilized red phosphorus in, 49
Swiss Army type, 45
three categories of, 46
three-component rimfire type, 56
triple salts, 48
two-component, 55
U.S. Cartridge Company NRA, 45
volume of gases, 42
Winchester 35-NF, 44
Priming powder, 15
Projectiles, 67
    baton rounds, 87
    bullets, 67–75
    chemical aspects, 67
    Dardick Trounds, 87
    exploding bullets, 78–80
    flare loads, 80
    flechette cartridges, 87
    frangible bullets, 87
    hand loading, 77–78
    history of, 3, 19
    multiple loads, 81–82
    poisoned bullets, 84–86
    saboted subcaliber bullets, 81
    shot loads, 86
    shotgun pellets and slugs, 75–77
    special purpose ammunition types, 82–84
    teargas bullets, 87–88
    wax bullets, 80
    wood bullets, 80
Propellant analysis, 183–188
    double-based propellants, 194
    quantitative, 260–261, 261
    single-based propellants, 193
Propellants, 4, 9, 57, 181
    burning rate, 58
    calcium carbonates as neutralizer in, 61
    chemical aspects, 57–66
    colored taggants in, 61
    comparison in suspect ammunition, 169
    comparison of fired and unfired, 171
    contribution to discharge particle content, 179
    defined, 57
    double-based, 62, 65
    energy/weight/bulk relationships, 57
    forensic chemical comparison, 103
    general specifications, 57
    homogeneity of, 169–171
    inorganic additives, 61
    powdered metals in, 61
    single-based, 64
    smokeless, 14
    stabilizers in, 61
    surface coatings, 58
    thermal characteristics, 61
    typical shapes, 58
    use in non-firearm devices, 66
Pyrodex, 60
Pyrophoric metal alloys, 51
    lack of sensitivity to percussion, 51
    in priming compositions, 50

Q

Quadrupole mass spectrometer, 117
Quinoil immonium ion, 107

R

Raman spectroscopy, 114
Range estimates, and bullet wipe pattern, 173
Reagan, Ronald, assassination attempt with exploding bullets, 80
Recent nontoxic primers, 201
Recoil, 31
Reflections, reducing with surface coatings, 98
Reliability of deposition, 271
Reloading, 77
Reloading speed, 23
Remington Accelerator rounds, 81
Remington Kleanbore, 46
Repeating firearms, 31
    historical development, 30
Resin, in caseless ammunition, 93
Resorcinol, 62

Retention time, 271
Retention volume, 115
Revolvers, 6
    defined, 5–6
    primer cups for, 40
    use in crime, 5
Rheinische-Westphalische Sprengstoff AG (RWS)
    development of Sinoxyd primer, 46
    noncorrosive primers from, 45
Rifled bore weapons, 8
Rifled slugs, 76
Rifles, 7
    evolution from muskets, 31
    primer cups for, 40
    use in crime, 5
Rifling, 19
Rimfire cartridges, 26
    rare use in kneecappings, 200
Rimfire round, 9
Round nose bullets, 67
    designs, 68
Round nose lead (RNL) ammunition, 129
Round of ammunition, 9
RPG7 rocket launcher, 163–166
    discharge residue from, 164
RTW bullet, 72
Russetting process, 98
Rusting, from potassium chlorate, 43

S

Sabot slugs, 77
Saboted subcaliber bullets, 81
Saltpeter, 13, 14
    in black powder, 59
Sample preparation, 268
Sampling
    apparatus for, 244
    assumptions about contamination, 249
    cartridge cases, 141–142
    clothing, 241–242, 242–246
    clothing examination, 248–249
    discussion, 249–251
    FDR/explosive residue, 142
    skin, 241–242
Sampling kits, 247–248
    incident report form in, 250
Sampling materials, storage of, 234
Sampling process, 233

Scanning electron microscope, 139–140, 348
    components, 113
    detectability of mercury particles by, 210
    percentage mercury indetectable by, 217
    proportion of mercury detectable by, 214
    use in sampling, 241
Scanning electron microscope with elemental analysis capability (SEM-EDX), 111, 200
    analysis of spent cartridge cases, 189
Schwartz, Berthold, 13
Sealers, 91
    in blank ammunition, 96
Self-loading pistols, 7
Sensitizers, in priming compositions, 42
Shark sticks, 96
Shaw, Joshua, 17
Shell X-Ploder bullets, 84
Shok-Lock rounds, 77
Shot cups, 76
Shot loads, 86
Shotgun cartridges, composition of, 35
Shotgun pellets, 75–77
Shotgun round, 9
Shotguns, 7
    reloading mechanisms, 8
    single- vs. double-barreled, 7
    as smooth bore weapons, 8
    use in crime, 5
Shoulder guns, 5
Silicon, as indicative vs. unique particles, 143
Single-based propellants, 64, 203, 253
    analysis of, 193
    diphenylamine as most common stabilizer in, 190
    outdoor firings with, 262
    and terrorist incidents in Northern Ireland, 241
    unreliable detection of, 264
Single entrance wounds, 82
Single-shot pistols, 7
Sinoxyd primer, 46
    lead styphnate in, 47
    use in U.K., 53
Sintox primer, 55, 137, 201, 224
    identification criteria, 272
    study of discharge particles, 228
    use in crime, 139
Skin, sampling for FDR, 241–242
Slow match, 15

Slugs, 75–77
Smith & Wesson, 26
   Nyclad ammunition development, 223
   perfection of centerfire cartridges by, 26
Smoke, surrounding bullets, 105
Smokeless powder, fragments in FDR, 125
Smokeless propellants, 14, 39, 59, 60, 63
Smooth bore weapons, 8
Sniper rifle, 5
Sodium nitrate, 60
Sodium rhodizonate test, 173
   for lead, 172
Soft shot, 75
Solid combustion products, 60
Solid phase extraction (SPE) system, 254–255, 266–267
   average recovery of organic components from swab via, 256
Special loads, 77
Special purpose ammunition types, 82–84
Specificity, of bulk elemental analysis methods, 110
Spent bullets
   comparison microscopy, 103
   FAAS analysis of, 184–188
   SEM/EDX analysis of, 189
Spent cartridge cases, 208
   elemental analysis, 202
   mercury loss from, 220
   mercury remaining in, 211
Spherical particles, in FDR, 125
Splat multipurpose ammunition, 84
Springs, construction materials, 97
Staba, 50
Stabanate, 49
Stabilized red phosphorus, in priming mixtures, 49
Stabilizers
   diphenylamine, 61, 62
   for nitrocellulose, 61
   in propellants, 61
   in single- and double-based propellants, 190
Starting pistols
   discharge residue, 146
   hardened steel blockage in barrel, 149
Steel
   grades of, 97
   use in cartridge cases, 35
Steel shot, 75
Stopping power, 81, 83–84
Storage areas, for sampling materials, 234
Stud guns, 143
Stun gun, 3–4
Styphnic acid, 55
Submachine guns, 5, 6
Suction sampling, 242, 250
   of clothing, 264
   comparison with swabbing, 243, 244, 248
Suction sampling device, 243
   revised, 246, 247
Sulfur, 14
   in black powder, 59
   in discharge residue particles, 166
   in gunpowder, 13
Supporting evidence, 273
Surface coatings, 98–100, 170–171
   of propellants, 58
Suspect processing procedures, 231
Suspect sampling time, 250
Suspects
   exclusion by absence of FDR, 272
   time of apprehension, 177–178
Swabbing methods
   average recovery of organic components using, 256
   comparison with suction sampling, 243, 244, 248
   of hand sampling, 132
   recovery of inorganic residue using, 267
   recovery of organic residue using, 265–266
Swabbing solvent, effect of water on, 162
Swiss Army, noncorrosive primer use, 45
Synthetic wood stocks, in firearm construction, 97

T

Taser gun, 4
Tear gas bullets, 87–88
Temperatures, inside gun during discharge, 10
Terminal ballistics, defined, 4
Terrorist attacks
   bullet fragmentation during, 163
   careful planning of, 175
   in Northern Ireland, xi, 241
Thermal energy analyzer, 140–141
Thermal stability, in priming compositions, 49

Index

Thiocyanate-chlorate primers, 44
Three-component rimfire primers, 56
THV bullet, 72
Time delays, in FDR sampling, 233–234
Time of discharge, difficulties with mercury as indicator of, 220
Tin, as indication of mercury primer, 208
Tin compounds, in propellants, 62
Tissue disruption, 82
Titanium
    as calcium silicide replacement, 55
    in tracer bullets, 73
TNT, 49
Total metal jacketed bullets (TMJ), 69, 223
Toy caps, particle classification scheme, 149–150
Trace level, 139
Tracer bullets, 72–73
    forensic determination of, 106
Trajectory tracers, 72
Tres Haute Vitesse, 72
Tributylphosphate, 192
Triple salts, in priming mixtures, 48
Trophy bonded bear claw, 70
Trophy bonded Sledgehammer, 70
Troubles, the, xi
Tungsten carbide, in AP bullets, 71, 72
Two-component primer compositions, 55

U

Ultra-Shock ammunition, 82
Ultraviolet spectroscopy, 114
Unique particles, 143, 157, 263, 272
    in FDR *vs.* blank cartridge residues, 147
    in particle classification scheme, 154
    percentage occurrence of accompanying elements in, 152–153
    proportion of, 148, 149
United Kingdom
    common priming mixtures, 54–55
    shift to NCNM primers, 47
    Sinoxyd type primers in, 53
Unjacketed bullets, 21, 68
    copper and zinc residues from, 148
    lubricants for, 91
    particles on new ammunition, 158, 159
U.S. Cartridge Company, NRA primer composition, 34

V

Varnishes, 91
Velet Cartridge Company, 80
Velet exploding bullets, 80
Verey pistols, 150
Visiprep D-L vacuum chamber, 266

W

Wads, 76
Wallace, Jim, xix
Water
    effect in swabbing solvent, 162
    effects on FDR, 161–162
Water reduction additives, 62
Wax bullets, 80
Weapons, defined, 3
Weather conditions, ignition dependence on, 23
Western primer, 46
Wet clothing, low retrieval rate of FDR, 161
Wheellock, 15, 30
Whitworth, Joseph, 20
Winchester Repeating Arms Company, 35-NF primer composition, 44
Winchester Silver tip, 70
Winchester Staynless, 46
Windshield shattering bullets, 84
Wood bullets, 80
Woods, in firearm construction, 97
World War I, nonmercuric primers of, 44
World War II
    exploding bullet use, 78
    FA-70 primers during, 44

X

X-ray analysis methods, 109

Z

Zinc, as residues from unjacketed bullets, 148